Springer Natural Hazards

More information about this series at http://www.springer.com/series/10179

Yu Huang · Zili Dai · Weijie Zhang

Geo-disaster Modeling and Analysis: An SPH-based Approach

 Springer

Yu Huang
Zili Dai
Weijie Zhang
Department of Geotechnical Engineering
Tongji University
Shanghai
China

ISBN 978-3-662-44210-4   ISBN 978-3-662-44211-1   (eBook)
DOI 10.1007/978-3-662-44211-1

Library of Congress Control Number: 2014945260

Springer Heidelberg New York Dordrecht London

Printed on acid-free paper

Springer is part of Springer Science+Business Media (www.springer.com)

Preface

Geo-disasters accompanied by large deformation and flow failure of geomaterials are a regular occurrence around the world. These disasters include landslides, debris flows, dam-breaks, soil liquefaction, seepage damage, and dynamic erosion. The disasters seriously damage infrastructure, resulting in casualties and heavy economic losses. To reduce the damage, one of the priorities for governments and researchers is to determine the probability of geo-disaster occurrence, devise hazard maps, and take protective measures. Therefore, there is an urgent need to study disaster mechanisms and simulate kinematic characteristics of the geomaterials during disaster evolution, such as runout distance, velocity, and impact force.

With consideration of the devastating impacts of geo-disasters, this monograph has the goal of presenting a Smoothed Particle Hydrodynamics (SPH)-based approach to geo-disaster modeling and analysis, thereby satisfying the needs of geo-disaster prevention and control efforts.

The monograph was assembled by systematically screening, sorting, and categorizing the authors' research work on the theory of the SPH method and its applications to geo-disasters. The basic SPH principle and algorithm are discussed, calculation programs based on elasto-plastic mechanics and fluid dynamics are coded, and visualization software is developed. Geo-disasters such as flow-like landslides, lateral flow of liquefied soil, and flow failures of solid waste in landfills are simulated and analyzed. Disaster evolution and dynamic characteristics are described. The monograph thereby provides a means for mapping hazardous areas, estimating hazard intensity, and identifying and designing appropriate protective measures.

The monograph has seven chapters, with Chap. 1 as the introduction. This describes certain catastrophic geo-disasters and their characteristics. Advantages of the SPH method are detailed and its applications to geo-disasters are reviewed.

Chapter 2 presents SPH theory and development history. Governing equations and constitutive laws are discretized based on SPH approximations. Important numerical techniques are introduced, such as the neighboring particle search algorithm, boundary treatment, and integration schemes. Then, SPH models for geo-disasters are established.

In Chap. 3, a computer procedure based on the SPH model is developed in the FORTRAN language environment, which is applicable to geo-disaster analysis and modeling. The function of each module in the procedure is explained via calculation flowcharts. Based on these calculation procedures, visual simulation software is developed through a Windows interface, and SPH computational efficiency is greatly improved.

In Chap. 4, a series of validations for the SPH numerical models are conducted. These include the dam-break model, soil flow model, simple shear, and soil nondrained shear tests. Results are compared with an analytical solution, test results from the literature, and solutions from other numerical methods.

The next three chapters focus on SPH applications to geo-disaster modeling and analysis, including municipal solid waste (MSW) flow slides in landfills (Chap. 5), lateral spreading of liquefied soil (Chap. 6), and flow-like landslides triggered by earthquakes (Chap. 7). In the above simulations, the dynamic behaviors of geomaterials are captured, such as flow velocity, runout distance, and impact force, and potential hazard areas are predicted. These are all useful for the prevention and control of geo-disasters.

The work uses the SPH method to study in detail the failure mechanism of geomaterials and geo-disaster formation and evolution. The reasons for selecting this method are as follows.

(1) SPH is one of the earliest mesh-free methods. It has clear advantages in dealing with problems of free surface, deformation boundary, motion interface, and extremely large deformation, which are involved in geo-disasters.

(2) Because of constant revision and improvement, the accuracy, stability, and adaptability of SPH have all met the requirements of practical project applications.

(3) SPH has a very wide range of applications, from continuous to discrete systems and from the micro-scale to macro-scale. It is even applicable to scales of astronomy. SPH has been successfully incorporated in commercial software packages and used in practical engineering.

The theoretical methods and related applications described in this monograph will benefit graduate students, engineers, researchers, and professionals in the fields of geologic, geotechnical, civil, and hydraulic engineering, as well as in computational mechanics. The monograph is simple to read. Background knowledge such as theories of the finite element method, finite difference method, and discrete element method are helpful but not necessary for readers to understand the theories and methods herein.

Our research group began SPH study in 2006 with the cooperation of Prof. Atsushi Yashima and Dr. Hideto Nonoyama (currently at Nagoya University, Japan) from the Department of Civil Engineering, Gifu University, Japan. We also received tremendous help from Prof. Kazuhide Sawada and Dr. Shuji Moriguchi (currently at Tohoku University, Japan). The authors would like to extend their heartfelt thanks to all of them.

After the 2008 Wenchuan earthquake in China, our group members traveled to the damage sites for survey. During the survey period, we had enthusiastic support from Prof. Qiang Xu's research group at Chengdu University of Technology, which the authors greatly acknowledge.

This monograph was completed with the joint efforts of all members in our research group at Tongji University. This group included graduate students Mr. Liang Hao, Mr. Pan Xie, and Mr. Chen Jin, who were involved in the comprehensive research work. Writing and editing were done with assistance from Ph.D. students Ms. Hualin Cheng, Mr. Chongqiang Zhu, and Ms. Yangjuan Bao.

The research work presented in the monograph was supported by the National Natural Science Foundation of China (Grant Nos. 40841014, 40802070, 41072202, 41111130205, 41211140042, 41372355), National Basic Research Program of China (973 Program) through Grant No. 2012CB719803, National Key Technologies R&D Program of China (Grant No. 2012BAJ11B04), State Key Laboratory of Geohazard Prevention and Geo-environment Protection (Grant No. DZKJ-0808), Key Laboratory of Engineering Geomechanics of the Chinese Academy of Sciences (Grant No. 2008-04), Shanghai Municipal Education Commission and Shanghai Education Development Foundation (Shu Guang Project No. 08SG22), Program for New Century Excellent Talents in University (Grant No. NCET-11-0382), Fundamental Research Funds for the Central Universities, and the Kwang-Hua Fund of the College of Civil Engineering at Tongji University. The authors express cordial acknowledgments to these organizations.

Although this monograph presents much of our research work, it represents only the first step in this field. Many relevant problems remain. All the authors hope that this monograph will attract more research interest in this area. If readers are interested in the basic program codes or visualization software, they may contact the first author, Prof. Yu Huang, at yhuang@tongji.edu.cn. Readers may use partial or full code at their own risk, as long as a proper reference and acknowledgment are given.

Because of our limited knowledge, there will be some inevitable omissions and mistakes in the monograph. Therefore, we welcome constructive criticism and correction toward continuously improving the work.

Shanghai, China
June 2014   Prof. Yu Huang

Contents

About the Authors

Yu Huang, first author of this monograph, Prof. Yu Huang, born 1973, received his Ph.D. in Geotechnical Engineering from Tongji University, Shanghai, China. He is now a deputy head of that department and a deputy director of the Key Laboratory of Geotechnical and Underground Engineering of the Ministry of Education at Tongji University.

Professor Huang's primary area of research includes geologic disasters, computational geomechanics, earthquake geotechnical engineering, environmental geology, and foundation engineering. He has authored more than 150 technical publications, including more than 30 papers in international refereed journals such as the *Bulletin of Engineering Geology and the Environment, Engineering Geology, Landslides, Natural Hazards, Environmental Earth Sciences, Geotextiles and Geomembranes,* and *Waste Management and Research.* He now serves on the editorial board for the *Bulletin of Engineering Geology and the Environment* (the official journal of IAEG), *Geotechnical Research* (ICE) and *Geoenvironmental Disasters* (Springer).

Co-author Zili Dai, born 1987, is a Ph.D. student of Prof. Yu Huang at Tongji University. He received his bachelor's degree in Civil Engineering from Shanghai University in 2010. He worked at Northwestern University and the University of California, Berkeley as a visiting scholar between 2012 and 2013.

Co-author Weijie Zhang, born 1986, was awarded bachelor's and master's degrees in Geological Engineering from Tongji University. He worked as a master's student in Prof. Yu Huang's research group between 2009 and 2012. He is currently a Ph.D. student in Nagoya Institute of Technology in Japan.

Figures

Tables

Chapter 1
Introduction

Geologic disaster is one of several types of adverse geologic conditions capable of causing damage or loss of property and life. It includes earthquakes, landslides, debris flow, soil liquefaction, rock falls, avalanches, tsunamis, and flooding. These disasters may be induced by natural factors and human activities, and can cause serious casualties and huge economic losses. Every year millions of people all over the world experience the effects of geologic disasters. New methods of predicting and preventing such events appear to be helpful, but are nowhere near perfect. This chapter treats the destruction of recent geo-disasters in the world and basic characteristics of large deformation in those disasters. Based on a review of studies on large deformation simulation and its current limitations, a novel mesh-free particle method called Smoothed Particle Hydrodynamics is introduced, and its advantages and disadvantages are detailed. The main content of the monograph and its innovation are also summarized here.

1.1 Geo-disasters and Analysis

Landslides are a continuing problem along hillsides, shorelines, and roadways, and represent a global issue since they occur worldwide. Among all landslides, flow-like ones are often catastrophic. Distinctive features of such landslides are extremely long travel distances and high sliding velocities. Earthquakes are one of the most important causes of these landslides. For example, the 2008 Wenchuan earthquake induced numerous landslides, collapses, unstable slopes, and other secondary geologic disasters. Of all the types of landslides caused by earthquakes, flow-like ones such as Tangjiashan, Wangjiayan, and Donghekou were the most significant. This is attributable to the extremely large volumes of displaced material (usually greater than 10 million m^3), high sliding velocities (in the order of meters per second), and long runout distances (from several hundred meters to a few kilometers). As a result, these flow-like landslides have caused numerous casualties and catastrophic destruction of buildings and regional landscapes (Huang and Li 2009).

© Springer-Verlag Berlin Heidelberg 2014
Y. Huang et al., *Geo-disaster Modeling and Analysis: An SPH-based Approach*,
Springer Natural Hazards, DOI 10.1007/978-3-662-44211-1_1

In addition to the flow-like landslides, laterally spreading liquefied soil can cause large deformation and flow failure. In geology, soil liquefaction refers to the process by which water-saturated, unconsolidated sediments are transformed into a substance that acts like a liquid, often during an earthquake. Under shearing or vibration during an earthquake, excess pore water pressure in the soil increases rapidly with the loss of grain contact and decreasing porosity. The soil then liquefies and behaves as a liquid. Hence, the strength and shear modulus of the soil decrease significantly after soil liquefaction, reducing the bearing capacity. By undermining foundations and base courses of infrastructure, this liquefaction can cause serious damage.

Seismic liquefaction has occurred in almost every strong earthquake. For example, a wide range of liquefaction produced serious disasters with the Xingtai, Haicheng, and Tangshan earthquakes of 1966–1976 (Wang et al. 1982; Chen et al. 1999; Yin et al. 2005). In the 2008 Wenchuan earthquake, the scope of liquefaction was concentrated in an area 160 long and 60 km wide. The greatest concentration was in an intensity VIII area. The wide-ranging flow of liquefied soil can damage structures and infrastructure, generally resulting in uneven settlement and cracks in the building superstructures in a liquefied area. For example, in an Ms 7.5 earthquake in Niigata, Japan during 1964, the foundation of an apartment was liquefied and the large deformation caused an overall tilt of the building (Kawasumi 1968). During the Tangshan earthquake of 1977, soil liquefaction led to substantial horizontal displacement and great destruction of highways and bridges. Dozens of buildings in a towel factory and steel factory in Tianjin cracked and collapsed (Chen 2001). In 1989, an Ms 7.1 earthquake occurred in Loma Prieta, California, USA. Earthquake-triggered liquefaction caused some revetment foundation slipping of more than 20 m in San Francisco (Bartlet and Youd 1995). Port facilities of the artificial island Port Island were damaged by liquefied foundation lateral flow during the Kobe earthquake of January 1995. On March 11, 2011, there was an earthquake of Mw 9.0 off the Pacific Coast of Tohoku, triggering widespread liquefaction in the Tohoku and Kanto regions of Japan. Many structures were damaged, such as in the widespread liquefaction around the parking area of Disneyland, and around the liquefaction flow slide of a pier in Miyagi Prefecture. According to the earthquake damage investigation, deformation was caused by earthquake-triggered soil liquefaction. In particular, the large deformation and lateral flow were the main reasons for damage to foundations, roads, bridges, dam slopes, buildings, and lifeline engineering structures. In view of this, research on flow deformation characteristics of sandy soil after seismic liquefaction is urgently needed.

In geologic disasters such as landslides, collapse, debris flow, unstable slopes, and sandy soil liquefaction, the deformation and failure of geomaterials are highly nonlinear. In such cases, geomaterial is in plastic or viscous mechanical states. The deformation is no longer in line with small deformation theory, but is in a nonlinear, large deformation flow state.

The phenomena of large deformation and flow failure of geomaterial not only occur in natural disasters but also in many human activities. Among the latter, soil excavation is an important cause of geomaterial deformation and failure. For example, slope excavation generated soil flow near China's Tianshengqiao hydropower station on December 24, 1985 (Chen 2003). Soil filling is also an important cause of geomaterial flow. The construction of dams and embankments on saturated soft ground often produces foundation sliding, such as soil flow on a weak foundation at Mahu Lake and failure of a railway foundation in Lianyungang. Foundation pit excavation can also cause substantial geomaterial deformation and instability. Improper foundation pits may instigate great losses, such as building foundation instability near Shanghai's Guangdong Road on September 1, 1994 and foundation pit instability of a Shijiazhuang shopping center in 1993. Construction of subway tunnels can also cause large deformation of geomaterials. For example, the tunnel between South Pudong Road and Nanpu Bridge of Shanghai Subway Metro Line No. 4 collapsed in July 2003, collapsing the surrounding buildings and causing land subsidence and flood prevention-wall cracking.

Given the geologic disasters and engineering accidents listed above, large geomaterial deformation and flow failure are very common, causing substantial casualties and economic loss. Therefore, study of large deformation and flow failure of geomaterials is of great significance to construction and disaster prevention and control.

The basic characteristics of such large deformation can be briefly summarized as follows:

(1) The deformation of geomaterials varies from tens of meters to several kilometers.
(2) Ground horizontal displacement caused by large soil deformation can lead to bridge length contraction and beam drop, broken piles in the foundation, displacement and tilt of piers, ruptures of underground pipelines, and displacements or cracks in roads and railways, which damage the structures or make them unusable.
(3) The large deformation of geotechnical materials is always triggered by a sudden increase of pore water pressure in soil or no pore water, such as in landslides and the collapse of dry sand or loess.
(4) Some large deformation analysis is within the scope of fluid dynamics theory.

1.2 Mesh-free Methods

Grid-based numerical methods such as the finite element method (FEM) have been powerful tools for engineering numerical analysis since the twentieth century, and have solved a large number of major scientific and engineering problems. However, FEM still has difficulty dealing with problems such as dynamic expansion of cracks, metal stamping forming, high-speed impact, damage and failure of material, corrosion and penetration, and large deformation problems of discontinuity. The method may have numerical difficulties (e.g., severe mesh winding, twisting, and distortion) when it comes to extremely large deformations. Consequently, remeshing is needed during the solution, which generally increases the complexity of computer programs and reduces their calculation precision. To avoid these numerical problems, mesh-free methods have begun to be developed in recent years. Compared with the grid-based method, these methods are based on point-based approximation. This can completely or partially eliminate the grid, obviating the need for initial grid division and reconstruction. In this way, the mesh-free method has great advantage for analysis of large deformation with respect to accuracy and efficiency. The basic concept is that a series of randomly distributed nodes (or particles) are used in lieu of a grid to solve a variety of integral equation or partial differential equations (PDEs) with different boundary conditions, thereby obtaining accurate and stable numerical solutions. These nodes or particles do not need to be connected by a grid.

However, for the large deformation of soil, mesh-free methods have limited applicability. Examples of this include the discrete element method (DEM), Discontinuous Deformation Analysis (DDA), element-free Galerkin (EFG), and cellular automata (CA). The reason is that nonphysical parameters in these methods are not confirmed accurately; moreover, they cannot describe the stress–strain relationship exactly.

Smoothed Particle Hydrodynamics (SPH), introduced and described in this monograph, is a mesh-free method based on pure Lagrangian description. This method can import a variety of material constitutive equations to describe mechanical properties. The basic concept of the method is to discretize continuous flows into a series of arbitrarily distributed particles carrying field variables, and then to analyze the mechanical behavior of the problem domain by analyzing particle trajectories and their interactions. SPH was first invented to solve astrophysical problems in three-dimensional open space by Lucy (1977) and Gingold and Monaghan (1977). Since the collective movement of those particles is similar to that of liquid or gas flow, it can be modeled by the governing equations of classical Newtonian hydrodynamics.

Advantages of the SPH method over traditional grid-based numerical methods are:

(1) Because of the Lagrangian property, the motion of the SPH particles can be traced and features of the entire physical system can be easily obtained. Therefore, it is much easier to identify free surfaces, moving interfaces, and deformable boundaries than with Eulerian methods. Time histories of the field variables (such as velocity and displacement) at each material point can also be obtained from the simulation. In addition, as a Lagrangian method, the SPH code is conceptually simpler than grid-based methods and should be faster, because there is no convective term in the related partial differential equations.

(2) Given the mesh-free character, an object under consideration in the SPH model can be discretized into a series of particles without use of a grid/mesh. Therefore, compared with grid-based methods, this distinctive mesh-free feature can process larger local distortion, because connectivity between the particles

can change with time. This feature has been used in many applications of solid mechanics, such as underground explosions, metal forming, high-velocity impact, crack growth, and fragmentation.

(3) As a particle method, discretization of complex geometry is simpler, because only an initial discretization is required. The refinement of particles is much easier to achieve than mesh refinement.

(4) SPH guarantees conservation of mass without extra computation, because the particles themselves represent mass. Pressure is computed from the weighted contributions of neighboring particles rather than by solving linear systems of equations.

(5) SPH is suitable for problems in which the material is not a continuum, and is therefore a valuable tool for numerical simulation of problems in bio- and nano-engineering at micro and nanoscale.

Although the favorable features of the SPH method and its applications to geotechnical engineering have been noted, there are drawbacks. These include inconsistency, tensile instability, and zero-energy modes, as described below.

(1) Inconsistency

The SPH method in its continuous form is inconsistent near boundaries, because of incomplete kernel support. Morris (1996) and Belytschko et al. (1996) identified a particle inconsistency problem that can lead to low accuracy in the SPH solution. Various solutions have been proposed to restore consistency and improve the accuracy of SPH. Randles and Libersky (1996, 2000) used the inconsistency to approximate the smoothing function and its derivatives to offset the inconsistency in approximating the field function and its derivatives. Vignjevic et al. (2000) implemented kernel normalization and correction in the corrected normalized smooth particle hydrodynamics (CNSPH) method, which is first-order consistent. These proposed modifications are based on either the kernel approximation or particle approximation. Recently, Chen and Beraun (2000) presented a corrective smoothed particle method (CSPM) based on Taylor series expansion of the SPH approximation of a function. Liu et al. (2003) improved the CSPM in discontinuous SPH (DSPH) methods to resolve problems with discontinuity, such as in shock waves. Other notable modifications or corrections of the SPH method that ensure first-order consistency include the EFG (Belytschko et al. 1994), reproducing kernel particle method (RKPM; Liu et al. 1995), moving least squares particle hydrodynamics (MLSPH) (Dilts 1999), and meshless local Petrov-Galerkin (MLPG) (Atluri and Zhu 2000). These methods allow restoration of consistency of any order by means of a correction function.

(2) Tensile instability

Tensile instability is a numerical problem that appears in the conventional SPH method and greatly limits its application in geological engineering. This instability manifests itself as a clustering of particles when they are under tensile stress. Swegle et al. (1995) first revealed this phenomenon and provided a stability criterion, noting that the tensile instability is closely related to the second derivative of the smoothing kernel function. Various remedies were proposed for this problem. Dyka and Ingel (1995), Dyka et al. (1997) added a series of additional stress points in the support domain other than the normal particles in a one-dimensional algorithm to remove the tensile instability. Randles and Libersky (2000, 2005) used these stress points to stagger all SPH particles, and extended this approach to multi-dimensional space. Monaghan (2000) removed the tensile instability by introducing an artificial stress. This method has been widely applied in geological engineering problems (Bui et al. 2008; Das and Cleary 2010; Karekal et al. 2011). Bonet and Kulasegaram (2001) discussed the problem of tension instability with the SPH method, stating that this instability is a property of a continuum where the stress tensor is isotropic and pressure is a function of density. They demonstrated that a stable solution can be obtained using Lagrangian CSPH without need of any artificial viscosity. More recently, a new method to avoid tensile instability was presented, with two sets of master and slave nodes used (Blanc and Pastor 2012b, 2013).

(3) Zero-energy modes

Zero-energy modes, first identified in an SPH solution by Swegle et al. (1994), represent modes of deformation characterized by a pattern of nodal displacement that produces zero strain energy. This problem can produce spurious oscillations and degrade the solution. The main cause of this condition is that all field variables and their derivatives are calculated at the same locations, so an alternating field variable has zero gradient at the particles. Two types of solutions are found in the literature: dissipation of spurious modes, and an alternative discretization that does not evaluate the variables and their derivatives at the same points. For example, an artificial stress was used to preclude instability (Randles and Libersky 2000). Two different sets of particles were used to evaluate stresses and velocities at separate points (Vignjevic et al. 2000). In addition, a stabilized updated Lagrangian formulation was incorporated in the SPH model to overcome the problem of zero-energy modes (Vidal et al. 2007).

Besides the major shortcomings discussed above, the SPH method has other defects. For example, the conventional method can only be used to simulate compressible fluid, a problem that was solved by the weakly compressible SPH (WCSPH) and incompressible (ISPH) methods, which are discussed in detail in Sect. 3.2. Another well-known problem is that the amount of smoothing needed for stability may dampen the short-wavelength structure (Hicks and Liebrock 1999) and smooth out strong shocks. This problem could be detrimental in analysis of certain geophysical processes that involve shock waves. Reformulation of standard SPH arithmetic for strong shock simulation has been proposed by Monaghan (1997), Inutsuka (2002), and Cha and Whitworth (2003). The common feature of these methods is a combination of standard SPH and Riemann solvers. Recently, Sigalotti et al. (2009) presented an adaptive SPH (ASPH) method for strong shocks. This method relied on an adaptive density kernel estimation (ADKE) algorithm, which allows the smoothing length to vary locally in space and time so that the minimum necessary smoothing is applied in regions of low density.

1.3 SPH Applications in Geo-disaster Modeling

In its early stage, application of SPH was mainly within the fields of astrophysics and hydrodynamics. In astrophysics, SPH was used to explain complicated problems, such as stellar collisions (Benz 1988; Monaghan 1992; Frederic and James 1999), supernovae (Hultman and Pharayn 1999), formation and collapse of galaxies (Monaghan and Lattanzio 1991; Berczik and Kolesnik (1993, 1998), (Berczik 2000), black hole coalescence with neutron stars (Lee 1998, 2000), single and multiple detonation of white dwarfs (Senz et al. 1999), and even evolution of the universe (Monaghan 1990). In hydrodynamics, SPH applications have included elastic flow, magnetic fluid dynamics, multiphase flow, quasi-incompressible flow, gravity flow in porous media, thermal conductivity, impact simulation, heat transfer, mass flow, and others.

The SPH method has recently been extended to a wide range of problems in both fluid and solid mechanics, with benefits from its strong ability to incorporate complex physical concepts into SPH formulations. A variety of SPH models have been proposed and applied to specific topics in geo-disasters, including dam breaks and coastal engineering, flow-like landslides, lateral spread of liquefied soil, seepage failure, dynamic erosion, underground explosions, and rock breakage. The feasibility and reliability of such models were verified successfully through comparisons with laboratory experiments, analytical solutions, and simulations with other methods. In the following section, some recent SPH applications to geo-disasters are described.

1.3.1 Dam-Breaks and Coastal Engineering

Given their successful applications in hydrodynamics, most SPH simulation studies of geo-disaster topics have concentrated on fields related to fluid dynamics, such as dam breaks and coastal engineering.

After a dam failure, large amounts of water stored in the reservoir suddenly rush downstream, destroying trees, dikes, buildings, and bridges along the way. To minimize the human and financial toll of dam-break disasters, it is important to predict the effects of catastrophic dam-break floods. Owing to its mesh-free, Lagrangian, and particle nature, SPH has been widely applied in studies related to dam breaks. Many interrelated aspects of dam-break problems have been investigated. These include approaches to the treatment of free-surface flow (Koh et al. 2012; Chang et al. 2011) and boundary conditions (Ata and Soulaimani 2005; Crespo et al. 2007), three-dimensional assessments of dam-break disasters (Roubtsova and Kahawita 2006; Ferrari et al. 2010), differences between Newtonian and non-Newtonian flow features (Shao and Lo 2003), interface analysis of multiphase flow (Colagrossi and Landrini 2003), and multiphase models for highly erosive flow (Shakibaeinia and Jin 2011). In these simulations, the unique advantages of SPH in dealing with free-surface and moving boundary problems were fully embodied. A pioneering work of SPH application to dam-break analysis was Wang and Shen (1999), who simulated inviscid dam-break flows and conducted depth-averaged analyses. A feature of their model is that the length of discrete parcels varies with changing flow conditions, so it is robust and especially suited to solving problems with sharp moving fronts.

Large sea waves occasionally break through coastal defenses and travel inland over long distances, resulting in damage to infrastructure and loss of life. Therefore, an important aspect of any mitigation effort is to predict the processes of wave generation, shoreward propagation, shoreline arrival, increase of water height, and wave breaking. Successful application of SPH methods to dam breaks has provided a foundation for the solution of fluid–structure interaction problems in coastal engineering.

A numerical model within the framework of an SPH method was established to analyze wave overtopping of ships and offshore platforms (Gomez-Gesteira et al. 2005). Some complex phenomena were successfully reproduced, including: (1) initial continuous flow; (2) flow separation after striking the deck; (3) varying wave behaviors above and below the deck; (4) formation of a jet after overtopping; and (5) flow restoration. All simulations accurately matched experimental observations. Interaction between waves and engineered coastal structures were studied using SPH by Mutsuda et al. (2008). In their model, the solid interface was automatically identified by particles overlapping fixed grids. The deformation and failure process of coastal structures subjected to impact from sea waves was calculated with smoothness, efficiency, and accuracy. A large eddy simulation (LES) approach was coupled with SPH to study the mechanics of near-shore solitary waves and address the typical problem of a solitary wave rising and falling against a vertical wall (Lo and Shao 2002). Following this, a similar SPH-LES model combined with a sub-particle-scale (SPS) turbulence model was presented for the treatment of turbulence associated with wave breaking (Shao et al. 2006). Configurations and overtopping characteristics of different types of waves (e.g., velocity fields and turbulent eddy viscosity distributions) were predicted. More recently, the friction force was added to the Navier-Stokes equations within an SPH framework to examine flow friction in porous media and the interface between waves and a breakwater covered by a layer of geomaterial (Shao 2010). This model has been validated as accurate for simulation of damped solitary and periodic waves over a porous bed, and applied to a breaking wave running up and over a breakwater protected by a porous geomaterial layer.

Using these SPH models, some significant problems in coastal engineering have been addressed, including free-surface, moving boundary, and solid-liquid coupling. Phenomena of dam break, wave dynamics (wave generation, breaking, and interaction with structures), and failure of breakwaters and their foundations were accurately reproduced and analyzed. The results provide a significant foundation for the design of offshore structures and the assessment of dam-break or tsunami disasters.

1.3.2 Slope Failure and Landslides

Natural slopes in soil and soft rock have become more vulnerable as human activities gradually extend into mountainous regions. Understanding the failure mechanism and post-failure behavior of slopes is very important for determining potential risk areas and devising hazard maps. The SPH method is considered an effective tool for modeling slope failure and predicting runout distance. For example, Bui et al. (2008) presented an SPH framework for stability analysis of a slope with reinforcing piles. They proposed an algorithm to deal with the problem of soil–structure interaction, which uses a coupling condition at the interface between soil and structure associated with a penalty force applied to different material particles near that interface. This model can simulate the following phenomena and analyze the mechanism in the following four cases: (1) development of shear bands by investigating an accumulated plastic strain contour plot; (2) the gross discontinuity of soil after failure; (3) stress distribution on the reinforcing pile; and (4) the bending mechanism of the reinforcing pile. The highlight of this research is the proposition of a well-performing contact algorithm to treat soil–structure interactions, which have constituted an enduring computing problem in geotechnical-related fields. This was the first known study implementing an elasto-plastic soil constitutive model (the Drucker-Prager model with a non-associated flow rule) into an SPH model to describe plastic soil behavior. However, particle deficiency near the solid boundary may be one of the most difficult problems for the SPH method when it is used to simulate elasto-plastic material. Although some solutions have been proposed, the low precision caused by lack of particle coverage near the solid boundary in this method requires further work.

Flow-like landslides often result in catastrophic events because of their relatively long runout distances and high velocities. A prediction of the runout, velocity, and impact force of flow-like landslides is necessary for adequate protective measures and risk management. A numerical model for dynamic analysis of rapid landslide motion across 3D terrain has been developed (McDougall and Hungr 2004). Depth-integrated equations were used to govern the mass and momentum balance of a column of earth material moving with the landslide. The model was tested using an analytical solution of the classical dam-break problem and a series of laboratory experiments. The model was then used to analyze the Frank Slide in Canada, with promising results. The model has many unique features, such as the ability to account for nonhydrostatic and anisotropic internal stress states, material entrainment along the slide path, and rheology variation. Its path material entrainment algorithm was described in detail by McDougall and Hungr (2005), and the importance of this capability was demonstrated using a back-analysis of the 1999 Nomash River landslide in Canada. Since then, depth-integrated models have been frequently used to model flow-like landslides. Of particular note is the pioneering work of Pastor et al. (2009). They proposed a depth-integrated model in combination with the SPH method to simulate propagation of flow-like catastrophic landslides. A velocity-pressure version of the Biot–Zienkiewicz model was introduced to consider the coupling effect between solid and fluid phases. As an example, the propagation stage of the catastrophic May 1998 landslide in the Tuostolo Basin in the Campania region of southern Italy was simulated. The model results were found to coincide with available field data. This model was later used to simulate the propagation stage of a 2001 lahar at the Popocatépetl volcano of Mexico (Haddad et al. 2010). The trajectory, velocities, depths, and runout distances of fluidized materials were correctly predicted. The sensitivity of the proposed model to rheological parameters was studied. The results showed that viscosity had a strong influence on flow velocity, whereas yield strength mainly affected runout distance.

Rock avalanches pose serious hazards to growing populations in mountainous areas. Sosio et al. (2012) investigated the evolving mobility of rock and debris avalanches in glacial environments using the SPH model proposed by McDougall (2006). The propagation outline, flow velocity, erosion depth, and deposit thickness were simulated. A quantitative comparison between the simulation results and field data was conducted, resulting in good agreement. Moreover, values of calibrated parameters were provided through back analyses.

Landslides, either submarine or aerial, can generate surface water waves that can cause damage and loss of life in coastal areas. Predicting the extent of these waves is important in flooding assessment. It remains difficult to simulate surface waves generated by a landslide, because of the complex motion of an underwater slope and its interaction with water. Because of its advantages in simulating free surfaces, moving boundaries, and large deformations, SPH is widely used to tackle the problem of landslide impulsive waves. For example, an SPH model was presented to simulate the 1958 Lituya Bay rockslide and resulting tsunami in Alaska (Schwaiger and Higman 2007). In this model, rock and water were treated as viscous and inviscid fluids, respectively, and the effect of air was neglected. A 2D SPH model for inviscid fluid was used to simulate landslide-induced waves and predict propagation of the water with satisfactory results (Qiu 2008). Numerical simulations of tsunami wave generation were carried out by Das et al. (2009). Complex flow patterns predicted in terms of free-surface profiles, shoreline evolution, and velocity fields were in good agreement with experimental data. A rheological SPH model was described by Capone et al. (2010) to investigate solid–liquid interaction, reproducing the generation and propagation of a tsunami triggered by underwater landslides.

To enhance stability and accuracy, the conventional SPH method used to simulate an arbitrarily moving compressible fluid was extended to incompressible or nearly incompressible flow, using two different approaches. The first approach is the WCSPH method, in which fluids are regarded as compressible with a sound speed that is much higher than bulk flow speed. A stiff equation of state is used to calculate pressure of the particles (Monaghan 1994). This method was corrected and used to simulate an impulsive wave generated by an underwater landslide (Ataie-Ashtiani and Mansour-Rezaei 2009). The method is easy to program because the pressure is obtained directly from an algebraic thermodynamic equation (Monaghan 1994). Nevertheless, there are some drawbacks. First, the artificial compressibility can cause problems with sound wave reflection at the boundary area (Shao and Lo 2003). Second, the time step is limited, because the sound speed is much greater than the maximum velocity (Lee et al. 2008). These problems can be overcome using a second approach, the ISPH. This method solves governing equations using prediction-correction fractional steps. Pressure is no longer a dependent variable, but can be computed from a pressure Poisson equation that satisfies the incompressibility condition. The main advantages of ISPH lie in its easy and efficient tracking of the free surface and the ease with which it treats wall boundaries. For example, Shao and Lo (2003) presented an ISPH method that was tested with a dam-break problem for Newtonian and non-Newtonian flows. The results were in good agreement with experimental data. An ISPH model was presented and tested using solitary waves generated by a heavy box falling into water (Ataie-Ashtiani and Shobeyri 2008). A submerged rigid wedge sliding along an inclined surface was simulated with this model. The computational ISPH results were in good agreement with experimental data. In addition, the proposed method was used to simulate the flow of gravel mass sliding along an inclined plane, accurately capturing wave profiles. Recently, a similar fractional step technique first proposed by Chorin (1968) was incorporated in the SPH model, to deal with coupled problems in geomechanics (Blanc and Pastor 2011; Blanc and Pastor 2012a). Comparisons of the ISPH algorithm with the classical WCSPH method were presented by Lee et al. (2008), who showed that ISPH yield results were much more reliable than with WCSPH. More recently, Shadloo et al. (2012) published a comparative study of the WCSPH and ISPH methods that provided numerical solutions for fluid flow over a square obstacle. They indicated that WCSPH produced numerical results as accurate and reliable as those of ISPH. Szewc et al. (2012) made a thorough comparison of these two incompressibility treatments. Their results showed that ISPH suffered from density accumulation errors, so a correction algorithm was used to improve accuracy. To the best of our knowledge, there has been no direct comparison between the conventional SPH and the two incompressibility treatments; hence, advantages of ISPH and WCSPH over the conventional SPH method cannot be given.

Unlike traditional numerical methods based on solid mechanics, the aforementioned SPH models analyze the large deformation and post-failure behavior of slopes from a fluid mechanics standpoint, and provide a completely new and effective approach for runout prediction in addition to the empirical methods. In the above simulations, complex constitutive models of geomaterials were imported in the SPH framework, such as the Bingham fluid and Drucker-Prager models. A simpler semi-empirical approach based on the concept of "equivalent fluid" was used in a new SPH model by McDougall and Hungr (2004) for a landslide study. Landslide material was governed by a simple rheology, and its parameters were selected based on back-analysis of full-scale landslides. The incorporation of these complex constitutive models in the SPH framework promotes the application of this method to geo-disasters. Moreover, in the simulation of submarine landslides, the interactions of various fluids were taken into account and tracked.

1.3.3 Liquefaction

Liquefaction occurs mainly in loose saturated sands as a result of earthquakes and rainfall. Subsoil lateral spreading after liquefaction can cause major damage to underground structures and infrastructure. The SPH method can be used for extremely large deformation involved in the lateral spread of liquefied subsoil. For example, a 2D SPH-based numerical model was introduced in a fluid dynamic framework to analyze lateral spread induced by liquefaction (Naili et al. 2005a). The liquefied subsoil was considered a non-Newtonian fluid by means of a Bingham fluid model. Under this hypothesis, the soil is capable of resisting any shear less than a yield defined by the residual shear strength. The ability of the method to reproduce the free-surface shape and obtain a time history of flow velocities was validated through comparisons with "shake table" experiment results. Through application of this model, the relationship between the shape of the velocity time curve, liquefied layer thickness, and surface ground slope was investigated, thereby clarifying mechanisms involved in liquefaction-induced lateral spread. Later, the flow of liquefied soil around a model pile was simulated and the drag force applied by liquefied flow was calculated (Naili et al. 2005b). In this work, the pile was discretized by a series of particles exerting a Lennard-Jones potential on the surrounding medium. The liquefied soil was again assumed a Bingham fluid, and a bilinear model was introduced to consider the recovery of rigidity. The proposed model can reproduce the external configuration of soil after liquefaction, distribution of flow velocities, and strain and stress fields of liquefied soil around the pile. However, both the soil-structure interaction at the interface and deformation and failure processes of the pile require more attention in future research.

1.3.4 Seepage

Large-scale deformation and hydraulic collapse of the ground induced by water flow through the ground are significant in the destabilization of dam foundations during floods, liquefaction, and other catastrophic events. To analyze these phenomena, Japanese researchers introduced SPH as a way to combine both discrete and continuum techniques for an analysis of ground failure linked with seepage (Maeda and Sakai 2004). A soil-water-air-coupled SPH model was proposed to model progressive seepage failure in soil. In this model, the solid and fluids are distributed in different computing layers. To combine layers of different phases, mixture theory was used to calculate frictional body forces resulting from velocity differences between adjacent phases. The application of this method led to numerical simulation of seepage processes around a sheet pile. The evolution of air bubbles during seepage was reproduced and deformation and failure of the ground induced by those air bubbles were successfully predicted.

Some improvements to the SPH method have been proposed in the course of this research. For example, a new procedure for calculating density sums different materials from individual given phases, thereby making it suitable for problems related to interfaces between different geologic materials. Another highlight of this research is accounting for solid-water-air bubble interactions using SPH and explaining gas generation and air bubble blow-out, which are regularly observed in association with seepage failures.

1.3.5 Dynamic Erosion

Dynamic erosion is a process in which soil and rock are removed from the earth surface under the influence of external factors such as water flow, and are subsequently transported and deposited in other locations. Excessive erosion may produce problems such as desertification, decreased agricultural productivity from land degradation, and ecological collapse from loss of nutrient-rich upper soil layers. With its advantages in simulation of flows involving rapid and large displacements, free surfaces, and moving interfaces, SPH is an effective technique for numerical modeling of dynamic erosion. A visual hydraulic erosion model based on fully 3D water dynamics has been described by Kristof et al. (2009). A physically based erosion model was incorporated in this model. Boundary particles were used to treat interactions (e.g., friction, sediment erosion, and deposition) and mediate sediment exchange between the moving fluid and underlying terrain. Two numerical examples were simulated and analyzed, including lake water eroding away from the boundary and waterfalls eroding underlying terrain. The ability of the proposed model to simulate the erosion of dense, large, and sparse fluid was demonstrated. SPH simulations of hydraulic erosion were conducted to investigate hydraulic erosion features of dams, levees, and earth embankments in storms and floods (Chen et al. 2011). In these simulations, fluid behavior was modeled by the Navier–Stokes equation, whereas the terrain was represented as a segmented height field. The erosion function was introduced with a critical shear stress to define the minimum shear stress on soil particles by water flows that could result in erosion. This model was verified by comparing the simulation results with those of physical tests. The formation of a gulley on a levee when overtopped was reproduced and better understanding of the erosion process was obtained, providing valuable knowledge for levee engineers. More recently, Manenti et al. (2012) proposed an SPH-based numerical model for prediction of coupled water-sediment dynamics induced by rapid water flow in an artificial reservoir. In their model, both water and soil were assumed weakly compressible viscous fluids governed by Navier–Stokes equations. Two erosion criteria, based on the Mohr-Coulomb yield criterion and Shields theory, were introduced for description of the failure mechanism of sediments. This model was validated by comparing the numerical results with laboratory test data.

1.3.6 Underground Explosions

Blast loading from an accidental explosion, blast excavation, or weapon attacks can cause extremely large deformation of soil and rock, and seriously damage nearby buildings and structures. Therefore, the response of soil and subsurface structures subjected to explosion loading has attracted much interest in current protective engineering research. Because of the difficulty and expense of large-scale field explosion tests, numerical simulations have become the main tool in such assessments. However, realistic computation can be very difficult because of the extremely large deformation involved and the need for complex modeling of interactions between modeled explosive detonations and the soil and buried structures. This has encouraged research into applications of SPH methods in this field. A coupled SPH-FEM approach has been proposed to simulate the response of soil and buried structures to blast loading (Wang et al. 2005). Large-scale soil deformation processes near an explosive charge and the response of remaining low-deformation regions were reproduced by SPH and FEM solutions, respectively. The SPH particles and FEM elements were joined on the interface, as shown in Fig. 1.1. On that basis, the 2D model was developed into a 3D one (Lu et al. 2005), and different response features determined by the 2D and 3D models were compared. This showed that propagation of a blast wave around edges and corners is much more complicated in 3D models, thereby strongly influencing loading conditions and the response of the structure. More recently, this modeling has been used to analyze liquefaction mechanisms induced by shock waves and investigate the effect of soil liquefaction on surface structures (Wang et al. 2011). A three-phase soil model for shock loading was proposed, and a completely joined surface was used to model interface interactions between the explosive detonation, soil medium, and geologic structures. Time history plots of pore water stress were constructed and the extent of liquefaction areas was predicted, which again coincided

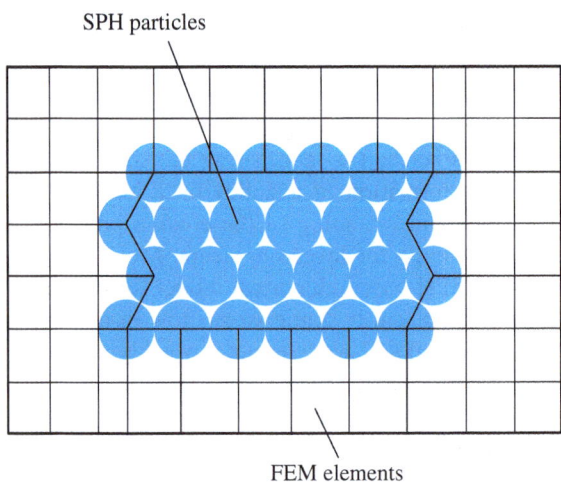

Fig. 1.1 Coupled mesh of SPH particles and FEM elements (based on Wang et al. 2005)

SPH particles

FEM elements

with empirical predictions. A conclusion was drawn, stating that surface structures would remain stable with limited permanent settlement if the liquefaction zone does not extend into the foundation region.

Xu and Liu (2008) reported on a coupled SPH-FEM model with a master–slave algorithm proposed by De Vuyst et al. (2005) that accounts for contact interaction of FE and SPH particles. The Jones-Wilkins-Lee (JWL) equation of state for high explosives was used in the blast analysis. The accuracy of this coupled approach and its advantage over the simulation of larger deformations were validated by free-field blast analysis. The modeling was used to simulate a subsurface blast, and the propagation of pressure and response of structures were reproduced.

It is concluded that SPH as applied to underground explosions has achieved significant outcomes. These include: (1) the coupled SPH-FEM approach to reproduce the dynamic response of soils and structures; (2) the three-phase soil modeling technique to simulate the internal distribution of stresses among soil components and describe changes of pore water pressure; (3) blast analysis based on an equation of state; (4) modeling of mechanisms of soil liquefaction caused by blast loading and the effect of liquefied soil on structures; and (5) stability criteria for surface structures subjected to underground explosions. These outcomes are important in the design of underground explosion protection systems and in minimizing damage caused by blast vibrations.

1.3.7 Rock Breakage

Rock caving is a large deformation problem that frequently results in the formation of crucial fractures and fragmentation. An SPH solution for elastic solid deformation was established in combination with a modified damage mode based on local stress history and flaw distribution. This solution was used to simulate rock caving processes (Karekal et al. 2011) and has had success in: (a) identifying rock deformation, fracture formation, and fragmentation processes that lead to progressive roof collapse; (b) characterizing the level of fracturing at a specific location; and (c) capturing the elastic-brittle and elasto-plastic material behavior of a rock mass. The coupled SPH-FEM simulations of rock caving as part of such research have proven to be computationally efficient.

Das and Cleary (2010) incorporated a continuum damage model into the SPH framework to simulate rock breakage under impact. Unconfined compressive strength was simulated as a validation of the SPH-based damage model for predicting rock fracture. Modeled and experimental results were in good agreement. Subsequently, the model was used to predict the brittle fracture of rock specimens of varying shapes during impact. This led to the conclusion that rock shape has a considerable influence on the fracture process, fragment sizes, energy dissipation, and post-fracture motion of fragments.

Mechanical characteristics and failure processes of heterogeneous rock-like materials were studied through SPH simulation of uniaxial and biaxial compression tests (Ma et al. 2011). Evolution of the failure was well captured. Cracks and fragments with large deformations were readily reproduced. The influences of material heterogeneity and confining load conditions were investigated.

The application of the SPH method to rock breakage reveals its superiority with regard to the simulation of extremely nonlinear physical processes such as fracture and fragmentation. The prediction of rock-breakage characteristics aids understanding of the fundamentals of rock failure and improves structural designs.

1.4 Monograph Outline

An SPH approach to geo-disaster modeling and analysis of the catastrophic damage caused by geo-disasters are introduced in this monograph. The concept and basic formulations of the SPH method are introduced. Corresponding numerical models

Fig. 1.2 Main logical structure of the monograph

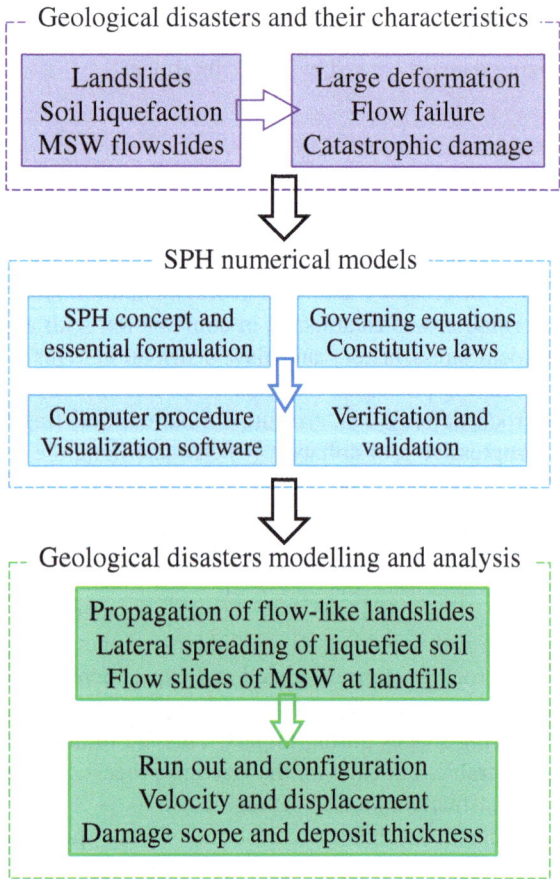

are established, as well as visualization simulation software. Different geomaterial constitutive laws are incorporated in the SPH formulations and the models are verified step-by-step. On these bases, the SPH models are used to analyze geo-disasters such as the lateral flow of liquefied soil, flow-like landslides, and flow slides of solid waste in landfills. Numerical results are compared with data measured onsite and obtained from model tests, achieving satisfactory consistency. The logical structure of this monograph is shown in Fig. 1.2.

References

Ata, R., & Soulaimani, A. (2005). A stabilized SPH method for inviscid shallow water flows. *International Journal for Numerical Methods in Fluids, 47*(2), 139–159.

Ataie-Ashtiani, B., & Mansour-Rezaei, S. (2009). Modification of weakly compressible smoothed particle hydrodynamics for preservation of angular momentum 813 in simulation of impulsive wave problems. *Coastal Engineering Journal, 51*(4), 363–386.

Ataie-Ashtiani, B., & Shobeyri, G. (2008). Numerical simulation of landslide impulsive waves by incompressible smoothed particle hydrodynamics. *International Journal for Numerical Method, 56*, 209–232.

Atluri, S., & Zhu, T. (2000). A new Meshless Local Petrov Galerkin (MLPG) approach in computational mechanics. *Computational Mechanics, 22*, 117–127.

Bartlet, S. F., & Youd, T. L. (1995). Empirical prediction of liquefaction-induced lateral spread. *Journal of Geotechnical Engineering, ASCE, 121*(4), 316–329.

Belytschko, T., Krongauz, Y., Organ, D., Fleming, M., & Krysl, P. (1996). Meshless methods: An overview and recent developments. *Computer Methods in Applied Mechanics and Engineering, 139*(1–4), 3–47.

Belytschko, T., Lu, Y. Y., & Gu, L. (1994). Element-free Galerkin methods. *International Journal for Numerical Methods in Engineering, 37*(2), 229–256.

Benz, W. (1988). Applications of smoothed particle hydrodynamics (SPH) to astrophysical problems. *Computer Physics Communications, 48*, 97–105.

Berczik, P. (2000). Modeling the star formation in galaxies using the chemodynamical SPH code. *Astronomy and Astrophysics, 360*, 76–84.

Berczik, P., & Kolesnik, I. G. (1993). Smoothed particle hydrodynamics and its applications to astrophysical problems. *Kinematics and Physics of Celestial Bodies, 9*, 1–11.

Berczik, P., & Kolesnik, I. G. (1998). Gas dynamical model of the triaxial protogalaxy collapse. *Astronomy and Astrophysical transactions, 16*, 163–185.

Blanc, T., & Pastor, M. (2011). Towards SPH modelling of failure problems in geomechanics a fractional step Taylor-SPH model. *European Journal of Environmental and Civil Engineering, 15*(SI), 31–49.

Blanc, T., & Pastor, M. (2012a). A stabilized fractional step, Runge-Kutta Taylor SPH algorithm for coupled problems in geomechanics. *Computer Methods in Applied Mechanics and Engineering, 221*, 41–53.

Blanc, T., & Pastor, M. (2012b). A stabilized Runge-Kutta, Taylor smoothed particle hydrodynamics algorithm for large deformation problems in dynamics. *International Journal for Numerical Methods in Engineering, 91*(13), 1427–1458.

Blanc, T., & Pastor, M. (2013). A stabilized smoothed particle hydrodynamics, Taylor-Galerkin algorithm for soil dynamics problems. *International Journal for Numerical and Analytical Methods in Geomechanics, 37*(1), 1–30.

Bonet, J., & Kulasegaram, S. (2001). Remarks on tension instability of Eulerian and Lagrangian corrected smooth particle hydrodynamics (CSPH) methods. *International Journal for Numerical Methods in Engineering 52*(11), 1203–1220

Bui, H. H., Sako, K., Fukagawa, R., Wells, J. C. (2008). SPH-based numerical simulations for large deformation of geomaterial considering soil-structure interaction. In *The 12th International Conference of International Association for Computer Methods and Advances in Geomechanics (IACMAG)* (pp. 1–6), Goa: India.

Capone, T., Panizzo, A., & Monaghan, J. J. (2010). SPH modeling of water waves generated by submarine landslides. *Journal of Hydraulic Research, 48*, 80–84.

Cha, S. H., & Whitworth, A. P. (2003). Implementation and tests of Godunov-type particle hydrodynamics. *Monthly Notices of the Royal Astronomical Society, 340*(1), 73–90.

Chang, T. J., Kao, H. M., Chang, K. H., & Hsu, M. H. (2011). Numerical simulation of shallow-water dam break flows in open channels using smoothed particle hydrodynamics. *Journal of Hydrology, 408*(1–2), 78–90.

Chen, J. K., & Beraun, J. E. (2000). A generalized smoothed particle hydrodynamics method for nonlinear dynamic problems. *Computer Methods in Applied Mechanics Engineering, 190*(1–2), 225–239.

Chen, W. H. (2001). Slipping disaster induced by seismic liquefaction. *Journal of Natural Disasters 10*(4), 88–93 (in Chinese).

Chen, W. H., Sun, J. P., & Xu, B. (1999). Recent development and trend in seismic liquefaction study. *World Information on Earthquake Engineering, 15*(1), 16–24. (in Chinese).

Chen, Z., Stuetzle, C. S., Cutler, B. M., Gross, J. A., Franklin, W. R., & Zimmie, T. F. (2011). *Analyses, simulations and physical modeling validation of levee and embankment erosion* (pp. 1503–1513). DallasUnited States: Geo-Frontiers 2011: Advances in Geotechnical Engineering.

Chen ZY (2003) *Soil slope stability analysis-theory, methods and programs.* Beijing: China Water and Power Press (in Chinese).

Chorin, A. J. (1968). Numerical solution of the Navier-Stokes equations. *Mathematics of Computation, 22*(104), 745–762.

Colagrossi, A., & Landrini, M. (2003). Numerical simulation of interfacial flows by smoothed particle hydrodynamics. *Journal of Computational Physics, 191*(2), 448–475.

Crespo, A. J. C., Gomez-Gesteira, M., & Dalrymple, R. A. (2007). Boundary conditions generated by dynamic particles in SPH methods. *CMC-Computers Materials and Continua, 5*(3), 173–184.

Das, K., Janetzke, R., Basu, D., Green, S., Stamatakos, J. (2009). Numerical simulations of Tsunami wave generation by submarine and aerial landslides using RANS and SPH. In *Models. 28th International Conference on Ocean, Offshore and Arctic Engineering, Honolulu, USA*, vol 5, pp. 581–594.

Das, R., & Cleary, P. W. (2010). Effect of rock shapes on brittle fracture using Smoothed Particle Hydrodynamics. *Theoretical and Applied Fracture Mechanics, 53*, 47–60.

De Vuyst, T., Vignjevic, R., & Campbell, J. C. (2005). Coupling between meshless and finite element method. *International Journal of Impact Engineering, 31*(8), 1054–1064.

Dilts, G. A. (1999). Moving-Least-Squares-particle hydrodynamics. I: Consistency and stability. *International Journal for Numerical Methods in Engineering, 44*(8), 1115–1155.

Dyka, C. T., & Ingel, R. P. (1995). An approach for tension instability in smoothed particle hydrodynamics (SPH). *Computers & Structures, 57*(4), 573–580.

Dyka, C. T., Randles, P. W., & Ingel, R. P. (1997). Stress points for tension instability in SPH. *International Journal for Numerical Methods in Engineering, 40*(13), 2325–2341.

Ferrari, A., Fraccarollo, L., Dumbser, M., Toro, E. F., & Armanini, A. (2010). Three-dimensional flow evolution after a dam break. *Journal of Fluid Mechanics, 663*, 456–477.

Frederic, A. R., & James, C. L. (1999). Smoothed particle hydrodynamics calculations of stellar interactions. *Journal of Computational and Applied Mathematics, 109*, 213–230.

Gingold, R. A., & Monaghan, J. J. (1977). Smoothed particle hydrodynamics: Theory and application to non-spherial stars. *Monthly Notices of the Royal Astronomical, 181*, 375–389.

Gomez-Gesteira, M., Cerqueiro, D., Crespo, C., & Dalrymple, R. A. (2005). Green water overtopping analyzed with a SPH model. *Ocean Engineering, 32*(2), 223–238.

Haddad, B., Pastor, M., Palacios, D., & Munoz-Salinas, E. (2010). A SPH depth integrated model for Popocatepetl 2001 lahar (Mexico): Sensitivity analysis and runout simulation. *Engineering Geology, 114*(3–4), 312–329.

Hicks, D. L., & Liebrock, L. M. (1999). SPH hydrocodes can be stabilized with shape-shifting. *Computers and Mathematics with Applications, 38*(5–6), 1–16.

Huang, R. Q., & Li, W. L. (2009). Analysis of the geo-hazards triggered by the 12 May 2008 Wenchuan Earthquake, China. *Bulletin of Engineering Geology and the Environment, 68*(3), 363–371.

Hultman, Pharayn A. (1999). Hierarchical, dissipative formation of elliptical galaxies: Is thermal instability the key mechanism hydrodynamic simulations including supernova feedback multi-phase gas and metal enrichment in cdm: Structure and dynamics of elliptical galaxies. *Astronomy & Astrophysics, 347*, 769–798.

Inutsuka, S. (2002). Reformulation of smoothed particle hydrodynamics with Riemann solver. *Journal of Computational Physics, 179*(1), 238–267.

Karekal, S., Das, R., Mosse, L., & Cleary, P. W. (2011). Application of a mesh-free continuum method for simulation of rock caving processes. *International Journal of Rock Mechanics and Mining Sciences, 48*, 703–711.

Kawasumi, H. (1968). *General report on the Niigata Earthquake of 1964*. Tokyo: Tokyo Electrical Engineering College Press.

Koh, C. G., Gao, M., & Luo, C. (2012). A new particle method for simulation of incompressible free surface flow problems. *International Journal for Numerical Methods in Engineering, 89*(12), 1582–1604.

Kristof, P., Benes, B., Krivanek, J., & St'ava, O. (2009). Hydraulic erosion using smoothed particle hydrodynamics. *Computer Graphics Forum, 28*(2), 219–228.

Lee, E. S., Moulinec, C., Xu, R., Violeau, D., Laurence, D., & Stansby, P. (2008). Comparisons of weakly compressible and truly incompressible algorithms for the SPH mesh free particle method. *Journal of Computational Physics, 227*, 8417–8436.

Lee, W. H. (1998). Newtonian hydrodynamics of the coalescence of black holes with neutron stars ii, tidally locked binaries with a soft equation of state. *Monthly Notices of the Royal Astronomical Society, 308*, 780–794.

Lee, W. H. (2000). Newtonian hydrodynamics of the coalescence of black holes with neutron stars iii, irrotational binaries with a stiff equation of state. *Monthly Notices of the Royal Astronomical Society, 318*, 606–624.

Liu, M. B., Liu, G. R., & Lam, K. Y. (2003). A one-dimensional meshfree particle formulation for simulating shock waves. *Shock Waves, 13*(3), 201–211.

Liu, W. K., Jun, S., & Zhang, Y. F. (1995). Reproducing kernel particle method. *International Journal for Numerical Methods in Fluids, 20*(8–9), 1081–1106.

Lo, E. Y. M., & Shao, S. D. (2002). Simulation of near-shore solitary wave mechanics by an incompressible SPH method. *Applied Ocean Research, 24*(5), 275–286.

Lu, Y., Wang, Z. Q., & Chong, K. (2005). A comparative study of buried structure in soil subjected to blast load using 2D and 3D numerical simulations. *Soil Dynamics and Earthquake Engineering, 25*, 275–288.

Lucy, L. B. (1977). A numerical approach to the testing of the fission hypothesis. *Astronomical Journal, 82*(12), 1013–1024.

Ma, G. W., Wang, X. J., & Ren, F. (2011). Numerical simulation of compressive failure of heterogeneous rock-like materials using SPH method. *International Journal of Rock Mechanics and Mining Sciences, 48*, 353–363.

Maeda, K., & Sakai, M. (2004). Development of seepage failure analysis procedure of granular ground with smoothed particle hydrodynamics (SPH) method. *Journal of Applied Mechanics JSCE, 7*, 775–786. (in Japanese).

Manenti, S., Sibilla, S., & Gallati, M. (2012). SPH simulation of sediment flushing induced by a rapid water flow. *Journal of Hydraulic Engineering, 138*(3), 272–284.

McDougall S (2006) A new continuum dynamic model for the analysis of extremely rapid landslide motion across complex 3d terrain. PhD thesis, University of British Columbia, Vancouver, Canada.

McDougall, S., & Hungr, O. (2004). A model for the analysis of rapid landslide motion across three-dimensional terrain. *Canadian Geotechnical Journal, 41*(6), 1084–1097.

McDougall, S., & Hungr, O. (2005). Dynamic modelling of entrainment in rapid landslides. *Canadian Geotechnical Journal, 42*(5), 1437–1448.

Monaghan, J. J. (1990). Modeling the universe. *Proceedings of the Astronomical Society of Australia, 18*, 233–237.

Monaghan, J. J. (1992). Smoothed particle hydrodynamics. *Annual Review of Astronomical and Astrophysics, 30*, 543–574.

Monaghan, J. J. (1994). Simulating free-surface flows with SPH. *Journal of Computational, 110*(2), 399–406.

Monaghan, J. J. (1997). SPH and Riemann solvers. *Journal of Computational Physics, 136*(2), 298–307.

Monaghan, J. J. (2000). SPH without a tensile instability. *Journal of Computational Physics, 159*(2), 290–311.

Monaghan, J. J., & Lattanzio, J. C. (1991). A simulation of the collapse and fragmentation of cooling molecular clouds. *Astrophysical Journal, 375*, 177–189.

Morris, J.P. (1996). Analysis of smoothed particle hydrodynamics with applications. Ph.D. Thesis, Monash University.

Mutsuda, H., Shinkura, Y,. Doi, Y. (2008). An eulerian scheme with lagrangian particles for solving impact pressure caused by wave breaking. In *18th International Offshore and Polar Engineering Conference* (Vol. 3, pp. 162–169). Vancouver, Canada.

Naili, M., Matsushima, T., & Yamada, Y. (2005a). A 2D smoothed particle hydrodynamics method for liquefaction induced lateral spreading analysis. *Journal of Applied Mechanics, 8*, 591–599.

Naili, M., Matsushima, T., Yamada, Y. (2005b) Smoothed particles hydrodynamics for numerical simulation of soil-structure problem due to liquefaction. In *Proceedings of the 40th Japan National Conference on Geotechnical Engineering, Hakodate*, pp. 2257–2258. (In Japanese).

Pastor, M., Haddad, B., Sorbino, G., & Drempetic, V. (2009). A depth-integrated, coupled SPH model for flow-like landslides and related phenomena. *International Journal for Numeracal and Analytical Methods in Geomechanics, 33*, 143–172.

Qiu, L. C. (2008). Two-dimensional SPH simulations of landslide-generated water waves. *Journal of Hydraulic Engineering, 134*(5), 668–671.

Randles, P. W., & Libersky, L. D. (1996). Smoothed particle hydrodynamics: some recent improvements and applications. *Computer Methods in Applied Mechanics and Engineering, 139*(1), 375–408.

Randles, P. W., & Libersky, L. D. (2000). Normalized SPH with stress points. *International Journal for Numerical Methods in Engineering, 48*(10), 1445–1462.

Randles, P. W., & Libersky, L. D. (2005). Boundary conditions for a dual particle method. *Computers & Structures, 83*(17–18), 1476–1486.

Roubtsova, V., & Kahawita, R. (2006). The SPH technique applied to free surface flows. *Computers & Fluids, 35*(10), 1359–1371.

Schwaiger, H. F., & Higman, B. (2007). Lagrangian hydrocode simulations of the 1958 Lituya Bay tsunamigenic rockslide. *Geochemistry Geophysics Geosysems, 8*(7), Q07006.

Senz, D. G., Bravo, E., & Woosley, S. E. (1999). Single and multiple detonations in white dwarfs. *Astronomy & Astrophysics, 349*, 177–188.

Sigalotti, L. D., Lopez, H., & Trujillo, L. (2009). An adaptive SPH method for strong shocks. *Journal of Computational Physics, 228*(16), 5888–5907.

Shadloo, M. S., Zainali, A., Yildiz, M., & Suleman, A. (2012). A robust weakly compressible SPH method and its comparison with an incompressible SPH. *International Journal for Numerical Methods in Engineering, 89*(8), 939–956.

Shakibaeinia, A., & Jin, Y. C. (2011). A mesh-free particle model for simulation of mobile-bed dam break. *Advances in Water Resources, 34*(6), 794–807.

Shao, S. D. (2010). Incompressible SPH flow model for wave interactions with porous media. *Coastal Engineering, 57*, 304–316.

Shao, S. D., Ji, C. M., Graham, D. I., Reeve, D. E., James, P. W., & Chadwick, A. J. (2006). Simulation of wave overtopping by an incompressible SPH model. *Coastal Engineering, 53*, 723–735.

Shao, S. D., & Lo, E. Y. M. (2003). Incompressible SPH method for simulating Newtonian and non-Newtonian flows with a free surface. *Advances in Water Resources, 26*(7), 787–800.

Sosio, R., Crosta, G. B., Chen, J. H., & Hungr, O. (2012). Modelling rock avalanche propagation onto glaciers. *Quaternary Science Reviews, 47*, 23–40.

Swegle, J. W., Hicks, D. L., & Attaway, S. W. (1995). Smoothed particle hydrodynamics stability analysis. *Journal of Computational Physics, 16*(1), 123–134.

Swegle, J. W., Attaway, S. W, Heinstein, M. W, Mello, F. J, Hicks, D. L. (1994). An analysis of smooth particle hydrodynamics. Sandia Report SAND93-2513.

Szewc, K., Pozorski, J., & Minier, J. P. (2012). Analysis of the incompressibility constraint in the smoothed particle hydrodynamics method. *International Journal for Numerical Methods in Engineering, 92*(4), 343–369.

Vidal, Y., Bonet, J., & Huerta, A. (2007). Stabilized updated Lagrangian corrected SPH for explicit dynamic problems. *International Journal for Numerical Method in Engineering, 69*(13), 2687–2710.

Vignjevic, R., Campbell, J., & Libersky, L. (2000). A treatment of zero-energy modes in the smoothed particle hydrodynamics method. *Computer Methods in Applied Mechanics and Engineering, 184*(1), 67–85.

Wang, K. L., Sheng, X. B., Chai, L. D., Hu, B. R., & Liu, H. M. (1982). Characteristics of liquation of soil in the areas with various intensities during Tangshan earthquake and criteria for recognition of liquation. *Seismology and Geology, 4*(2), 59–70. (in Chinese).

Wang, Z., & Shen, H. T. (1999). Lagrangian simulation of one-dimensional dam-break flow. *Journal of Hydraulic Engineering, 125*(11), 1217–1220.

Wang, Z. Q., Lu, Y., & Bai, C. (2011). Numerical simulation of explosion-induced soil liquefaction and its effect on surface structures. *Finite Elements in Analysis and Design, 47*, 1079–1090.

Wang, Z. Q., Lu, Y., Hao, H., & Chong, K. (2005). A full coupled numerical analysis approach for buried structures subjected to subsurface blast. *Computers & Structures, 83*(4–5), 339–356.

Xu, J. X., & Liu, X. L. (2008). Analysis of structural response under blast loads using the coupled SPH-FEM approach. *Journal of Zhejiang University Science A, 9*(9), 1184–1192.

Yin, R. Y., Liu, Y. M., Li, Y. L., & Zhang, S. M. (2005). The relation between earthquake liquefaction and landforms in Tangshan region. *Research of Soil and Water Conservation, 12*(4), 110–112. (in Chinese).

Chapter 2
SPH Models for Geo-disasters

The Smoothed Particle Hydrodynamics (SPH) method is one of the earliest mesh-free methods of pure Lagrangian description, and has been widely used in many fields of engineering. In this chapter, the development history, basic concept, and essential formulations of SPH are introduced. Governing equations and constitutive laws are incorporated into the SPH framework, and SPH models for geo-disaster modeling and analysis are established based on existing research (Liu and Liu 2003; Moriguchi 2005; Nonoyama 2011).

2.1 Basic Concept of SPH

SPH is a true mesh-free particle method based on a pure Lagrange description, which was first developed to solve astrophysical problems in three-dimensional open space, particularly polytopes (Lucy 1977; Gingold and Monaghan 1977). Later, Monaghan and Lattanzio (1985) summarized basic concepts of discretization for the governing equations, including continuity, momentum, and energy. Selection of the smoothing kernel function and techniques used in deriving SPH formulations for complex partial differential equations (PDEs) have been discussed. Considering the instability of the SPH numerical solution, Swegle et al. (1995), Dyka (1994), and Chen et al. (1999) proposed stabilization schemes. Johnson and Beissel (1996) put forth a calculation method for stress.

The core of this method is fully implied in the three words Smoothed Particle Hydrodynamics. "Smoothed" represents the smoothed approximation nature of using weighted averages of neighboring particles for stability. "Particle" indicates that the method is based on mesh-free particle theory. The computing domain is treated using a discrete particle instead of continuous entities. "Hydrodynamics" points to the fact that the SPH method was first applied to hydrodynamics problems.

The basic concept of SPH is that a continuous fluid is represented by a set of arbitrarily distributed particles. The moving particles possess material properties. By providing accurate and stable numerical solutions for hydrodynamic equations

© Springer-Verlag Berlin Heidelberg 2014
Y. Huang et al., *Geo-disaster Modeling and Analysis: An SPH-based Approach*,
Springer Natural Hazards, DOI 10.1007/978-3-662-44211-1_2

and tracking movements of each particle, the method can describe the mechanical behavior of an entire system. Therefore, the key facet of SPH is how to solve the PDEs using a series of arbitrarily distributed particles carrying field variables, such as mass, density, energy, and stress tensors. In an actual situation, it is usually difficult to obtain an analytical solution of these PDEs, which gives rise to the need for numerical methods for them. The first step is to discretize the problem domain of the PDEs. Then, there is a need to approximate the variable function and its derivative for the arbitrary particles. Finally, the approximate functions are applied to the PDEs to obtain a series of discretization Ordinary Differential Equations (ODEs), which are only related to time.

The core concept of the SPH method can be summarized as follows.

1. In the SPH model, the problem domain is replaced by a series of arbitrarily distributed particles. There is no connectivity between these particles, which reflects the mesh-free nature of this method. The major concern of this method is how to ensure the stability of numerical solutions, especially in applying the arbitrarily distributed particles to solve problems with derivative boundary conditions.
2. One of the most important steps is to represent a function in continuous form as an integral representation using an interpolation function. This step is usually called kernel approximation. The integral has a smoothing effect, similar to the weak form equations. In reality, the kernel approximation stabilizes numerical calculation of the SPH.
3. Another important step is that the value of a function at computing particle a is approximated using the averages of function values at all neighboring particles within the horizon of particle a. This step is termed particle approximation. The role of this approximation is to generate banded or sparse discretized system matrices, which are extremely important for calculation efficiency.
4. Using an explicit integration algorithm to solve differential equations can achieve fast time stepping. The time history of all field variables for all the particles can also be obtained. An appropriate method to determine the time step must be selected in the SPH method.

In summary, the mesh-free, adaptive, stable, and Lagrangian-description SPH method can be used as a dynamics problem solver.

2.2 SPH Approximation

The SPH method is built on interpolation theory, with two essential aspects. The first is smoothed (or kernel) approximation, which represents a function in continuous form as an integral representation. The other is particle approximation, which represents the problem domain using a set of discrete particles within the influence domain to estimate field variables for those computing particles. The value of a function at computing particle a is approximated using the average of those values

Fig. 2.1 SPH concept figure
(reprinted from Huang et al.
(2011) with permission of
Springer)

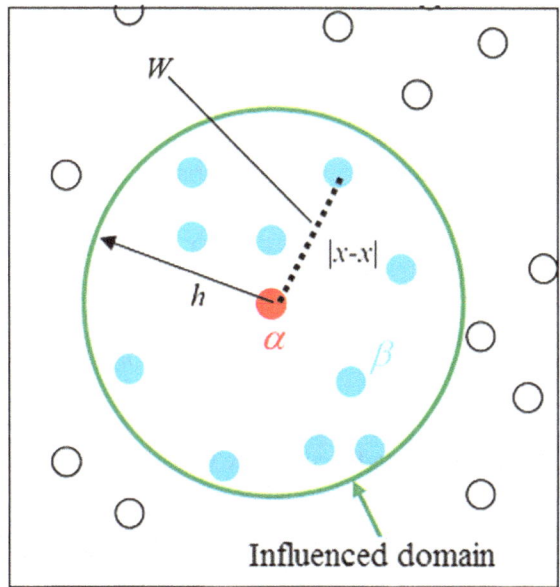

of the function at all particles in the influence domain of particle a, weighted by
the smoothing function. The radius of influence domain is defined as h, depending
on the precision of specific problems. As shown in Fig. 2.1, W is the smoothing
kernel function, α is the computing particle, and β is the neighboring particle.

2.2.1 Kernel Approximation

The integral representation of a function $f(x)$ used in the SPH method can be
rewritten in a continuous form:

$$f(x) = \int_D f(x')\,\delta(x - x')\,dx',\tag{2.1}$$

where $f(x)$ is a function of the three-dimensional position vector x, D is the volume
of the integral that contains x, and $\delta(x - x')$ is the Dirac delta function, given by

$$\delta(x - x') = \begin{cases} 1 & x = x' \\ 0 & x \neq x' \end{cases}.\tag{2.2}$$

If $f(x)$ is defined as a continuous function in D, Eq. (2.1) is exact or rigorous. If
the Delta function kernel $\delta(x - x')$ is replaced by a smoothing kernel function
$W(x - x', h)$, the integral representation of $f(x)$ is approximated by

$$f(x) \approx \int_D f(x')\, W(x - x',\, h)\, dx'. \tag{2.3}$$

The smoothing kernel function is also known as the interpolation kernel function, and has the following two characteristics:

$$\int_D W(x - x',\, h)\, dx' = 1, \tag{2.4}$$

$$\lim_{h \to 0} W(x - x',\, h) = \delta(x - x'). \tag{2.5}$$

W is usually chosen according to the above requirements, which should also be a differentiable even function. The Gaussian kernel, index kernel, cubic spline functions, B-spline function, and quartic spline functions are common smoothing kernel functions.

The above describes the kernel approximation, which is denoted by angle brackets $\langle \rangle$. Hence, Eq. (2.3) can be written as

$$\langle f(x) \rangle = \int_D f(x')\, W(x - x',\, h)\, dx'$$
$$= \int_D f(x')\, W(r,\, h)\, dx'. \tag{2.6}$$

It is obvious that Eq. (2.4) is a unity condition. Equation (2.5) considers the value of W at the point $x' = x$, and is a strongly peaked function. If $|x - x'| > h$, the value of W is zero. If h tends to zero, W is the delta function kernel $\delta(x - x')$ and it can therefore be given as $\lim_{h \to 0}\langle f(x) \rangle = f(x)$. The $\langle f(x) \rangle$ is a kernel approximation of $f(x)$, which can be considered as a smoothing or filter for $W(x - x',\, h)$, filtering local statistical fluctuation of $W(x - x',\, h)$.

Equation (2.6) is a standard expression of the kernel approximation of a function. The approximation of the spatial derivative $\nabla f(x)$ is obtained simply by replacing $f(x)$ with $\nabla f(x)$ in Eq. (2.6), which gives

$$\langle \nabla f(x) \rangle = \int_D \nabla f(x')\, W(x - x',\, h)\, dx'$$
$$= \int_D \nabla f(x')\, W(r,\, h)\, dx'. \tag{2.7}$$

Because the above equation cannot be calculated directly, the following deformation is conducted:

$$\left[\nabla f(x')\right] W(x - x', h) = \nabla[f(x') W(x - x', h)] - f(x')\nabla W(x - x', h). \quad (2.8)$$

$$\langle\nabla f(x)\rangle = \int_D \nabla\left[f(x') W(x - x', h)\right] dx' - \int_D f(x')\nabla W(x - x', h) dx'$$
$$= \int_S f(x') W(x - x', h) \, \boldsymbol{n} dS - \int_D f(x')\nabla W(x - x', h) dx', \quad (2.9)$$

where S is the boundary surface of the integral volume and $\boldsymbol{n}$ is the unit vector normal to surface S. Because of the compact support condition of W, the surface integral of Eq. (2.5) is zero in the influence domain D. Then Eq. (2.9) can be rewritten as

$$\langle\nabla f(x)\rangle \approx - \int_D f(x') \nabla W(x - x', h) dx'. \quad (2.10)$$

Equation (2.10) is the most commonly used form of function approximation for $\nabla f(x)$.

2.2.2 Particle Approximation

If kernel function W is an n-time differentiable function, it can be deduced that $\langle f(x)\rangle$ is also such a function.

Assuming that the fluid is flowing with density $\rho(x)$, it is divided into N volume elements. Masses of the volume elements are m_1, m_2, m_3,..., m_N, respectively, and positions of corresponding centers of mass are x_1, x_2, x_3,..., x_N, respectively. The continuous SPH integral representation for $f(x)$ can be written in the following form of discretized particle approximation:

$$\langle f(x)\rangle = \sum_{j=1}^{N} m_j \frac{f_j(x')}{\rho_j} W(x - x', h). \quad (2.11)$$

Based on Eq. (2.11), Eqs. (2.6) and (2.10) can be written as

$$\langle f_i(x)\rangle = \sum_{j=1}^{N} \frac{m_j}{\rho_j} f_j(x') W(x - x', h), \quad (2.12)$$

$$\langle\nabla f_i(x)\rangle = - \sum_{j=1}^{N} \frac{m_j}{\rho_j} f_j(x')\nabla W(x - x', h). \quad (2.13)$$

Following the same argument, the particle approximation for the spatial derivative of the function is

$$
\begin{aligned}
\frac{\partial \langle f(x) \rangle}{\partial x_i} &= \frac{\partial}{\partial x_i} \int f_j(x') \, W(x - x', \, h) \mathrm{d}x' \\
&= \int f_j(x') \frac{\partial W(x - x', \, h)}{\partial x_i} \mathrm{d}x' \\
&= \sum_{j=1}^{N} \frac{m_j}{\rho_j} f_j(x') \frac{\partial W(x - x', \, h)}{\partial x_i}.
\end{aligned}
\tag{2.14}
$$

Here, we use $\rho(x)$ instead of $f(x)$. From Eq. (2.11), $\rho(x)$ can be obtained as

$$
\langle \rho_i(x) \rangle = \sum_{j=1}^{N} m_j W(x - x', \, h).
\tag{2.15}
$$

This states that the density of particles can be approximated by the weighted average of particle densities within the influence domain of a particle. This is the meaning of the "smoothed particle."

2.3 SPH Model for Hydrodynamics

2.3.1 Governing Equations

Navier–Stokes (N–S) equations are introduced in this section as the governing equations to describe the incompressible Newtonian fluid.

2.3.1.1 Equation of Continuity

The continuity equation is based on the conservation of mass, which is obtained as

$$
\frac{\mathrm{d}\rho}{\mathrm{d}t} = -\rho \frac{\partial u^{\beta}}{\partial x^{\beta}},
\tag{2.16}
$$

where ρ is density, t is time, u represents the velocity vector, and the superscript β denotes the coordinate directions (as does the parameter α in subsequent equations). The above equation can be transformed as follows:

$$
\frac{\mathrm{d}\rho}{\mathrm{d}t} = -\frac{\partial \left(\rho u^{\beta} \right)}{\partial x^{\beta}} + u^{\beta} \frac{\partial \rho}{\partial x^{\beta}}.
\tag{2.17}
$$

2.3.1.2 Equation of Motion

In general, the equation of motion is given by

$$\frac{\mathrm{d}u^\alpha}{\mathrm{d}t} = \frac{1}{\rho}\frac{\partial \sigma^{\alpha\beta}}{\partial x^\beta} + F, \tag{2.18}$$

where σ represents the stress tensor and F the external force vector. The first term on the right side of Eq. (2.18) can be expanded as follows:

$$\frac{1}{\rho}\frac{\partial \sigma^{\alpha\beta}}{\partial x^\beta} = \frac{\partial}{\partial x^\beta}\left(\frac{\sigma^{\alpha\beta}}{\rho}\right) + \frac{\sigma^{\alpha\beta}}{\rho^2}\frac{\partial \rho}{\partial x^\beta}. \tag{2.19}$$

2.3.1.3 Equation of Energy

The equation of energy is based on the conservation of energy, which is a representation of the first law of thermodynamics.

$$\frac{\partial e}{\partial t} = \frac{\sigma^{\alpha\beta}}{\rho}\frac{\partial v^\alpha}{\partial x^\beta}. \tag{2.20}$$

2.3.2 SPH Formulation for Governing Equations

Using the smoothing approximation and particle approximation mentioned in the last section, the SPH version of the governing equations can be expressed as follows:

$$\text{Equation of continuity}: \quad \frac{\mathrm{d}\rho_i}{\mathrm{d}t} = \sum_{j=1}^{N} m_j\left(u_i^\beta - u_j^\beta\right)\frac{\partial W_{ij}}{\partial x_i^\beta}, \tag{2.21}$$

$$\text{Equation of motion}: \quad \frac{\mathrm{d}u_i^\alpha}{\mathrm{d}t} = \sum_{j=1}^{N} m_j\left[\frac{\sigma_i^{\alpha\beta}}{(\rho_i)^2} + \frac{\sigma_j^{\alpha\beta}}{(\rho_j)^2}\right]\frac{\partial W_{ij}}{\partial x_j^\beta} + F_i, \tag{2.22}$$

$$\text{Equation of energy}: \quad \frac{\mathrm{D}e_i}{\mathrm{D}t} = \frac{1}{2}\sum_{j=1}^{N} m_j\left(\frac{P_i + P_j}{\rho_i\rho_j}\right)v_{ij}^\beta\frac{\partial W_{ij}}{\partial x_i^\beta} + \frac{\mu_i}{2\rho_i}\varepsilon_i^{\alpha\beta}\varepsilon_j^{\alpha\beta}, \tag{2.23}$$

where W_{ij} is the smoothing function of particle i evaluated at particle j, and is closely related to h.

$$W_{ij} = W(x_i - x_j, h) = W(R_{ij}, h), \tag{2.24}$$

where $R_{ij} = r_{ij}/h = |x_i - x_j|/h$ is the relative distance of particles i and j, and r_{ij} is the distance between these two particles.

In this monograph, we consider the temperature of soil material as a constant. Therefore, the energy in Eq. (2.23) is not considered.

2.3.3 Newton Fluid Constitutive Model

The Newton fluid stress tensor is

$$\sigma_{ij} = -P\delta_{ij} + \xi D_{kk}\delta_{ij} + 2\eta D_{ij}, \tag{2.25}$$

where

$$D_{ij} = \frac{1}{2}\left(\frac{\partial u_i}{\partial x_j} + \frac{\partial u_j}{\partial x_i}\right), \tag{2.26}$$

where σ_{ij} is a stress tensor, P is pressure, δ_{ij} is the Kronecker delta function, ξ is the second viscosity coefficient, η is the viscosity coefficient, and D_{ij} is a shear strain tensor.

Introducing the Stokes hypothesis

$$\xi + \frac{2}{3}\eta = 0. \tag{2.27}$$

Then Eq. (2.25) can be rewritten as

$$\sigma_{ij} = -P\delta_{ij} - \frac{2}{3}\eta D_{ij}\delta_{ij} + 2\eta D_{ij}. \tag{2.28}$$

2.3.4 SPH Formulation for Poisson's Equation

To solve the N–S equations used in this monograph, we should first obtain the pressure by solving Poisson's equation. Together with velocity u_k, pressure p_k, and external force F_k, we introduce a virtual velocity u^* to construct the equation. Then the N–S equations may be obtained.

From Eqs. (2.18), (2.26) and (2.28), the incompressible N–S governing equation for the momentum equation becomes

$$\frac{Du_i}{Dt} = -\frac{1}{\rho}\frac{\partial P}{\partial x_j} + \frac{\eta}{\rho}\frac{\partial}{\partial x_j}\left(\frac{\partial u_i}{\partial x_j}\right) + F_i. \tag{2.29}$$

That is,

$$\frac{Du}{Dt} = -\frac{1}{\rho}\mathrm{grad}P + \nu\nabla^2 u + F. \tag{2.30}$$

Here, we introduce u^*, which is defined as

$$u^* = u^k + \Delta t\left(-\frac{1}{\rho}\mathrm{grad}P^k + \nu\nabla^2 u^k + F^k\right). \tag{2.31}$$

Simultaneously, we assume u^k and P^k as

$$u^{k+1} = u^k + u', \tag{2.32}$$

$$P^{k+1} = P^k + P'. \tag{2.33}$$

Adding Eq. (2.31) to (2.32), we obtain the following:

$$
\begin{aligned}
\left(u_i^\alpha\right)^{k+1} &= \left(u_i^\alpha\right)^k + \Delta t\left(-\frac{1}{\rho}\operatorname{grad}\left(P^\alpha\right)^{k+1} + \nu\nabla^2\left(u_i^\alpha\right)^k + F^k\right)\\
&= \left(u_i^\alpha\right)^* - \Delta t\left(-\frac{1}{\rho}\operatorname{grad}\left(P^\alpha\right)^k + \nu\nabla^2\left(u_i^\alpha\right)^k + F^k\right)\\
&\quad + \Delta t\left(-\frac{1}{\rho}\operatorname{grad}\left(P^\alpha\right)^{k+1} + \nu\nabla^2\left(u_i^\alpha\right)^k + F^k\right)\\
&= \left(u_i^\alpha\right)^* + \Delta t\frac{1}{\rho}\left(\operatorname{grad}\left(P^\alpha\right)^k - \operatorname{grad}\left(P^\alpha\right)^{k+1}\right),
\end{aligned}
\tag{2.34}
$$

$$
\begin{aligned}
u_i^{\alpha\,k+1} &= u_i^{\alpha\,k} + \Delta t\left(-\frac{1}{\rho}\frac{\partial P^{k+1}}{\partial x_i} + \eta\frac{\partial^2 u_i^{\alpha\,k}}{\partial x_j^2} + F^k\right)\\
&= u_i^{\alpha*} - \Delta t\frac{1}{\rho}\frac{\partial P'}{\partial x_i},
\end{aligned}
\tag{2.35}
$$

where α is the computing particle in the center and β represents neighboring particles within the influence domain.

The derivative of Eq. (2.35) is

$$
\frac{\partial\left(u_i^\alpha\right)^{k+1}}{\partial x_i} = \frac{\partial\left(u_i^\alpha\right)^*}{\partial x_i} - \Delta t\frac{\partial}{\partial x_i}\left(\frac{1}{\rho^\alpha}\nabla\left(P'^\alpha\right)\right).
\tag{2.36}
$$

We note that

$$
\frac{\partial\left(u_i^\alpha\right)^{k+1}}{\partial x_i} = \operatorname{div}\left(u_i^\alpha\right)^{k+1} = 0.
\tag{2.37}
$$

Consequently, Eq. (2.36) can be rewritten as

$$
\frac{\partial\left(u_i^\alpha\right)^*}{\partial x_i} = \Delta t\frac{\partial}{\partial x_i}\left(\frac{1}{\rho^\alpha}\nabla\left(P'^\alpha\right)\right).
\tag{2.38}
$$

The continuity equation gives

$$
\frac{D\rho}{Dt} + \rho_0\nabla u = 0.
\tag{2.39}
$$

Because fluid density ρ is proportional to the number of particles n, we can obtain

$$
\frac{1}{n^0}\frac{Dn}{Dt} + \nabla\cdot u = 0.
\tag{2.40}
$$

Here, we introduce n^*. From the definition of the smoothed function, we can obtain

$$
n^* = \sum_{\alpha\neq\beta}^{N} W^{\alpha\beta},
\tag{2.41}
$$

and

$$\frac{1}{n^0}\frac{n^0 - n^*}{\Delta t} + \nabla \cdot u = 0. \tag{2.42}$$

The n^0 is obtained by

$$n^0 = n^{k+1} = n^* + n'. \tag{2.43}$$

From simultaneous Eqs. (2.38), (2.42), and (2.43), we can obtain

$$\nabla u = -\frac{n^* - n^0}{n^0 \Delta t} = \Delta t \frac{\partial}{\partial x_i}\left(\frac{1}{\rho^\alpha}\nabla\left(P'^\alpha\right)\right). \tag{2.44}$$

Then the Poisson's equation is obtained by:

$$\nabla\left(\frac{1}{\rho^\alpha}\nabla\left(P'^\alpha\right)^{k+1}\right) = -\frac{n^* - n^0}{n^0 \Delta t^2}. \tag{2.45}$$

Referring to the SPH formulation for the energy equation proposed by Cleary and Monaghan (1999),

$$\frac{1}{\rho}\nabla \cdot (k\nabla T) = \frac{\partial U^\alpha}{\partial t} = \sum_{\beta}^{N}\frac{4m^\beta}{\rho^\alpha \rho^\beta}\frac{k^\alpha k^\beta}{k^\alpha + k^\beta}\frac{\left(T^\alpha - T^\beta\right)\left(x^\alpha - x^\beta\right)\cdot\frac{\partial W^{\alpha\beta}}{\partial x_i}}{\left(x^\alpha - x^\beta\right)^2 + \left(y^\alpha - y^\beta\right)^2}. \tag{2.46}$$

Applying the SPH discrete equations to the left side of Poisson's equation, we can obtain the following:

$$\nabla \cdot \left(\frac{1}{\rho^\alpha}\nabla\left(P'^\alpha\right)\right) = \sum_{\beta}^{N}\frac{4m^\beta}{\rho^\beta}\frac{\frac{1}{\rho^\alpha}\frac{1}{\rho^\beta}}{\frac{1}{\rho^\alpha} + \frac{1}{\rho^\beta}}\frac{\left(\left(P'^\alpha\right) - \left(P'^\beta\right)\right)\left(x_i^\alpha - x_i^\beta\right)\cdot\frac{\partial W^{\alpha\beta}}{\partial x_i}}{\left(x^\alpha - x^\beta\right)^2 + \left(y^\alpha - y^\beta\right)^2}$$

$$= \sum_{\beta}^{N}\frac{4m^\beta}{\rho^\beta}\frac{1}{\rho^\alpha + \rho^\beta}\frac{\left(\left(P'^\alpha\right) - \left(P'^\beta\right)\right)\left(x_i^\alpha - x_i^\beta\right)\cdot\frac{\partial W^{\alpha\beta}}{\partial x_i}}{\left(x^\alpha - x^\beta\right)^2 + \left(y^\alpha - y^\beta\right)^2}. \tag{2.47}$$

The two-dimensional Poisson equation is represented as follows:

$$\nabla \cdot \left(\frac{1}{\rho^\alpha}\nabla\left(P'^\alpha\right)\right) = \sum_{\beta}^{N}\frac{4m^\beta}{\rho^\beta}\frac{1}{\rho^\alpha + \rho^\beta}\frac{\left(\left(P'^\alpha\right) - \left(P'^\beta\right)\right)\left(x^\alpha - x^\beta\right)\cdot\frac{\partial W^{\alpha\beta}}{\partial x}}{\left(x^\alpha - x^\beta\right)^2 + \left(y^\alpha - y^\beta\right)^2}$$

$$+ \sum_{\beta}^{N}\frac{4m^\beta}{\rho^\beta}\frac{1}{\rho^\alpha + \rho^\beta}\frac{\left(\left(P'^\alpha\right) - \left(P'^\beta\right)\right)\left(y^\alpha - y^\beta\right)\cdot\frac{\partial W^{\alpha\beta}}{\partial y}}{\left(x^\alpha - x^\beta\right)^2 + \left(y^\alpha - y^\beta\right)^2}$$

$$= \sum_{\beta}^{N}\frac{4m^\beta}{\rho^\beta}\frac{1}{\rho^\alpha + \rho^\beta}\left(\frac{\left(x^\alpha - x^\beta\right)\cdot\frac{\partial W^{\alpha\beta}}{\partial x} + \left(y^\alpha - y^\beta\right)\cdot\frac{\partial W^{\alpha\beta}}{\partial y}}{\left(x^\alpha - x^\beta\right)^2 + \left(y^\alpha - y^\beta\right)^2}\right)\left[\left(P'^\alpha\right) - \left(P'^\beta\right)\right]. \tag{2.48}$$

$$-\frac{n^* - n^0}{n^0 \Delta t^2} = \sum_{\beta}^{N} \frac{4m^\beta}{\rho^\beta} \frac{1}{\rho^\alpha + \rho^\beta} \left(\frac{\left(x^\alpha - x^\beta\right) \cdot \frac{\partial W^{\alpha\beta}}{\partial x} + \left(y^\alpha - y^\beta\right) \cdot \frac{\partial W^{\alpha\beta}}{\partial y}}{\left(x^\alpha - x^\beta\right)^2 + \left(y^\alpha - y^\beta\right)^2} \right) \left(\left(P'^\alpha\right) - \left(P'^\beta\right) \right).$$

$$(2.49)$$

That is,

$$-\frac{n^* - n^0}{n^0 \Delta t^2} = \sum_{\beta}^{N} \frac{4m^\beta}{\rho^\beta} \frac{1}{\rho^\alpha + \rho^\beta} \left(\frac{\left(x^\alpha - x^\beta\right) \cdot \frac{\partial W^{\alpha\beta}}{\partial x} + \left(y^\alpha - y^\beta\right) \cdot \frac{\partial W^{\alpha\beta}}{\partial y}}{\left(x^\alpha - x^\beta\right)^2 + \left(y^\alpha - y^\beta\right)^2} \right) \left(P'^\alpha\right)$$

$$- \sum_{\beta}^{N} \frac{4m^\beta}{\rho^\beta} \frac{1}{\rho^\alpha + \rho^\beta} \left(\frac{\left(x^\alpha - x^\beta\right) \cdot \frac{\partial W^{\alpha\beta}}{\partial x} + \left(y^\alpha - y^\beta\right) \cdot \frac{\partial W^{\alpha\beta}}{\partial y}}{\left(x^\alpha - x^\beta\right)^2 + \left(y^\alpha - y^\beta\right)^2} \right) \left(P'^\beta\right).$$

$$(2.50)$$

The above equation can be solved by the incomplete Cholesky decomposition conjugate gradient (ICCG) method.

2.4 SPH Model for Solid Mechanics

Compared with other numerical methods, SPH has shown a distinct advantage in dealing with large deformation problems. The method has been widely applied in many disciplines, and has evolved from a hydrodynamics technique to a mechanics one. In this section, the Drucker–Prager material law has been incorporated into the SPH model to promote the application of that model in solid mechanics.

2.4.1 Governing Equations

2.4.1.1 SPH Approximation of Equation of Continuity

When SPH is used to calculate the elasto-plastic mechanic problems, there is still a need to rewrite the equation of continuity in the SPH formulations.

$$\rho_i = \sum_{j=1}^{N} m_j W_{ij}.$$

$$(2.51)$$

Similarly, Eq. (2.52) is used to treat the free-surface and material interface.

$$\rho_i = \frac{\sum_{j=1}^{N} m_j W_{ij}}{\sum_{j=1}^{N} \frac{m_j}{\rho_j} W_{ij}}.$$

$$(2.52)$$

Equation (2.53) is the mass conservation equation in a Euler description. In a Lagrangian description, it can be expressed as Eq. (2.54). Therefore, another SPH formulation for the equation of continuity is obtained as Eq. (2.55).

$$\frac{\partial \rho}{\partial t} + \frac{\partial (\rho v_i)}{\partial x_i} = 0, \tag{2.53}$$

$$\frac{\partial \rho}{\partial t} = -\frac{\partial (\rho v_i)}{\partial x_i} + v_i \frac{\partial \rho}{\partial x_i}, \tag{2.54}$$

$$\frac{d\rho}{dt} = -\sum_{\beta}^{N} \frac{m^{\beta}}{\rho^{\beta}} \rho^{\beta} v_i^{\beta} \frac{\partial W^{\alpha\beta}}{\partial x_i} + v_i \sum_{\beta}^{N} \frac{m^{\beta}}{\rho^{\beta}} \rho^{\beta} \frac{\partial W^{\alpha\beta}}{\partial x_i} = \sum_{\beta}^{N} m^{\beta} (v_i^{\alpha} - v_i^{\beta}) \frac{\partial W^{\alpha\beta}}{\partial x_i^{\beta}}, \tag{2.55}$$

2.4.1.2   SPH Approximation of Equation of Momentum

The equation of momentum in the elasto-plastic SPH model is similar to that in fluid dynamics.

$$\frac{dv_i}{dt} = \frac{1}{\rho} \frac{\partial \sigma_{ij}}{\partial x_j} + F_i, \tag{2.56}$$

where σ_{ij} is the stress tensor and F_i is body force.

Evaluating the first item on the right side of Eq. (2.56), then

$$\frac{\partial}{\partial x_i} \left(\frac{\sigma_{ij}}{\rho} \right) = \frac{\rho \frac{\partial \sigma_{ij}}{\partial x_i} - \sigma_{ij} \frac{\partial \rho}{\partial x_i}}{\rho^2} = \frac{\partial}{\partial x_i} \left(\frac{\sigma_{ij}}{\rho} \right) + \frac{\sigma_{ij} \frac{\partial \rho}{\partial x_i}}{\rho^2}. \tag{2.57}$$

Substituting Eq. (2.57) into (2.56), we have

$$\frac{dv_i}{dt} = \frac{\partial}{\partial x_i} \left(\frac{\sigma_{ij}}{\rho} \right) + \frac{\sigma_{ij} \frac{\partial \rho}{\partial x_i}}{\rho^2} + F_i. \tag{2.58}$$

Using the smoothing and particle approximations, the SPH formulations for the equation of momentum in the elasto-plastic SPH model can be described by

$$\begin{aligned}
\frac{Dv_i^{\alpha}}{Dt} &= \frac{\sigma_i^{\alpha\beta}}{\rho_i^2} \sum_{j=1}^{N} \frac{m_j}{\rho_j} \rho_j \frac{\partial W_{ij}}{\partial x_i^{\beta}} + \sum_{j=1}^{N} \frac{m_j}{\rho_j} \frac{\sigma_j^{\alpha\beta}}{\rho_j} \frac{\partial W_{ij}}{\partial x_i^{\beta}} + F_i \\
&= \sum_{j=1}^{N} m_j \left(\frac{\sigma_i^{\alpha\beta}}{\rho_i^2} + \frac{\sigma_j^{\alpha\beta}}{\rho_j^2} \right) \frac{\partial W_{ij}}{\partial x_i^{\beta}} + F_i.
\end{aligned} \tag{2.59}$$

An additional artificial viscosity is usually incorporated into the pressure terms of the equation of motion to convert kinetic energy to heat (Monaghan and Gingold 1983; see Eq. (2.60)). This avoids numerical oscillation and resists penetration between particles.

$$\frac{\mathrm{D}v_i^\alpha}{\mathrm{D}t} = \sum_{j=1}^{N} m_j \left(\frac{\sigma_i^{\alpha\beta}}{\rho_i^2} + \frac{\sigma_j^{\alpha\beta}}{\rho_j^2} - \delta^{\alpha\beta} \Pi_{ij} \right) \frac{\partial W_{ij}}{\partial x_i^\beta} + F_i. \tag{2.60}$$

This viscosity is defined by the following formulas:

$$\Pi_{ij} = \frac{-a\bar{c}\mu_{ij} + b(\mu_{ij})^2}{\overline{\rho^{\alpha\beta}}}, \tag{2.61}$$

$$\mu_{ij} = \frac{h(v_i - v_j)\sqrt{(x_i - x_j)^2 + (y_i - y_j)^2}}{(x_i - x_j)^2 + (y_i - y_j)^2 + kh^2}, \tag{2.62}$$

$$\bar{c} = \frac{1}{2}(c_i + c_j), \tag{2.63}$$

$$\bar{\rho} = \frac{1}{2}(\rho_i + \rho_j), \tag{2.64}$$

where a and b are constant parameters (often chosen to be 1.0) that improve numerical stability of the code (Monaghan 1988; Evard 1988). k is a free parameter used to avoid numerical divergence when particles approach each other, and can be taken as 0.01 (Liu and Liu 2003). c_i and c_j are the velocity of sound at points i and j, respectively.

2.4.2 Elasto-Plastic Constitutive Model

2.4.2.1 Stress–Strain Relation

For elasto-plastic materials, the constitutive model in the form of (Jaumann) stress rate can be used to establish a relationship between stress states and particle motion. To deal with the large deformation problem, total deformation can be separated into elastic and plastic deformations, but the former is relatively small.

From elastic mechanics, the stress-strain relationship of elastic deformation can be represented as

$$\sigma_{ij} = D_{ijkl}^e \varepsilon_{kl}^e, \tag{2.65}$$

where σ_{ij} is the elastic stress tensor, D_{ijkl}^e is a fourth-order elastic coefficient matrix, and ε_{kl}^e is elastic strain. The incremental form of the elastic stress–strain relationship can be represented as

$$\mathrm{d}\sigma_{ij} = D_{ijkl}^e \mathrm{d}\varepsilon_{kl}^e, \tag{2.66}$$

$$[D^e] = \begin{bmatrix} \lambda + 2\mu & \lambda & \lambda & 0 & 0 & 0 \\ \lambda & \lambda + 2\mu & \lambda & 0 & 0 & 0 \\ \lambda & \lambda & \lambda + 2\mu & 0 & 0 & 0 \\ 0 & 0 & 0 & \mu & 0 & 0 \\ 0 & 0 & 0 & 0 & \mu & 0 \\ 0 & 0 & 0 & 0 & 0 & \mu \end{bmatrix}, \tag{2.67}$$

where parameters λ and γ are

$$\lambda = \frac{\upsilon E}{(1 + \upsilon)(1 - 2\upsilon)}, \tag{2.68}$$

$$\mu = \frac{E}{2(1 + \upsilon)}, \tag{2.69}$$

where E is Young's modulus and υ is Poisson's ratio.

Combining Eqs. (2.66) and (2.67), the stress–strain matrix can be obtained as

$$\begin{bmatrix} \sigma_{xx} \\ \sigma_{yy} \\ \sigma_{zz} \\ \sigma_{xy} \\ \sigma_{yz} \\ \sigma_{zx} \end{bmatrix} = \begin{bmatrix} \lambda + 2\mu & \mu & \mu & 0 & 0 & 0 \\ \mu & \lambda + 2\mu & \mu & 0 & 0 & 0 \\ \mu & \mu & \lambda + 2\mu & 0 & 0 & 0 \\ 0 & 0 & 0 & \mu & 0 & 0 \\ 0 & 0 & 0 & 0 & \mu & 0 \\ 0 & 0 & 0 & 0 & 0 & \mu \end{bmatrix} \begin{bmatrix} \varepsilon^e_{xx} \\ \varepsilon^e_{yy} \\ \varepsilon^e_{zz} \\ \varepsilon^e_{xy} \\ \varepsilon^e_{yz} \\ \varepsilon^e_{zx} \end{bmatrix}. \tag{2.70}$$

In the plain strain problem, σ_{yz}, σ_{zx}, ε_{zz}, ε_{yz}, and ε_{zx} are all zero. Therefore, D^e_{ijkl} can be obtained as follows:

$$\begin{bmatrix} \sigma_{xx} \\ \sigma_{yy} \\ \sigma_{zz} \\ \sigma_{xy} \end{bmatrix} = \begin{bmatrix} \lambda + 2\mu & \mu & 0 \\ \mu & \lambda + 2\mu & 0 \\ \mu & \mu & 0 \\ 0 & 0 & \mu \end{bmatrix} \begin{bmatrix} \varepsilon^e_{xx} \\ \varepsilon^e_{yy} \\ \gamma^e_{xy} \end{bmatrix}. \tag{2.71}$$

It is well known that three problems must be considered to establish an elasto-plastic constitutive model. The first is the plastic potential function, which determines the direction of plastic strain. The second is the yield function, which determines whether plastic strain emerges. The last is the compatibility equation, which determines the magnitude of plastic strain.

Simple tensile test results show that the full yield strain ε_{ij} is composed of two parts, elastic strain ε^e and plastic strain ε^p:

$$\varepsilon_{ij} = \varepsilon^e_{ij} + \varepsilon^p_{ij}. \tag{2.72}$$

The differential of Eq. (2.72) is

$$d\varepsilon_{ij} = d\varepsilon^e_{ij} + d\varepsilon^p_{ij}. \tag{2.73}$$

It can be rewritten as

$$d\varepsilon_{ij}^e = d\varepsilon_{ij} - d\varepsilon_{ij}^p. \tag{2.74}$$

Substituting the above equation into Eq. (2.66), we can obtain

$$d\sigma_{ij} = D_{ijkl}^e \left(d\varepsilon_{ij} - d\varepsilon_{ij}^p \right). \tag{2.75}$$

The flow rule of plastic strain is represented as

$$d\varepsilon_{ij}^p = d\lambda \frac{\partial g}{\partial \sigma_{ij}} \quad (d\lambda \geq 0), \tag{2.76}$$

where $d\lambda$ depends on particle position and the load level. This represents the property of the plastic material and can determine the magnitude of plastic strain. g is the plastic potential function. When the yield surface coincides with the plastic potential surface, the yield function can act as the plastic potential function, which can be expressed as

$$\begin{cases} d\varepsilon_{ij}^p = d\lambda \dfrac{\partial f}{\partial \sigma_{ij}} \quad (d\lambda \geq 0). \\ f = g \end{cases} \tag{2.77}$$

Because of the plastic deformation, g is related not only to the stress state but also to the loading history. Here, a hardening parameter is introduced to represent this history. Then, the yield function f can be written as

$$f(\sigma_{ij}, L) = 0. \tag{2.78}$$

The total derivative of the above equation may be written as

$$df = \frac{\partial f}{\partial \sigma_{ij}} d\sigma_{ij} + \frac{\partial f}{\partial L} \frac{\partial L}{\partial \varepsilon_{ij}^p} d\varepsilon_{ij}^p = 0. \tag{2.79}$$

Substituting Eq. (2.75) into the above equation, we can obtain

$$\frac{\partial f}{\partial \sigma_{ij}} D_{ijkl}^e \left(d\varepsilon_{kl} - d\varepsilon_{kl}^p \right) + \frac{\partial f}{\partial L} \frac{\partial L}{\partial \varepsilon_{ij}^p} d\varepsilon_{ij}^p = 0. \tag{2.80}$$

This can be rewritten as

$$\frac{\partial f}{\partial \sigma_{ij}} D_{ijkl}^e d\varepsilon_{kl} = \left(\frac{\partial f}{\partial \sigma_{ij}} D_{ijkl}^e - \frac{\partial f}{\partial L} \frac{\partial L}{\partial \varepsilon_{kl}^p} \right) d\varepsilon_{kl}^p. \tag{2.81}$$

Substituting Eq. (2.77) into the above equation, we can obtain

$$\frac{\partial f}{\partial \sigma_{ij}} D_{ijkl}^e d\varepsilon_{kl} = \left(\frac{\partial f}{\partial \sigma_{ij}} D_{ijkl}^e - \frac{\partial f}{\partial L} \frac{\partial L}{\partial \varepsilon_{kl}^p} \right) d\lambda \frac{\partial f}{\partial \sigma_{kl}}, \tag{2.82}$$

where dλ can be solved as

$$d\lambda = \frac{\frac{\partial f}{\partial \sigma_{ij}} D^e_{ijkl} d\varepsilon_{kl}}{\left(\frac{\partial f}{\partial \sigma_{ij}} D^e_{ijkl} - \frac{\partial f}{\partial L} \frac{\partial L}{\partial \varepsilon^p_{kl}}\right) \frac{\partial f}{\partial \sigma_{kl}}}. \tag{2.83}$$

Substituting Eq. (2.77) into (2.75), it can be rewritten as

$$d\sigma_{ij} = D^e_{ijkl}\left(d\varepsilon_{kl} - d\lambda \frac{\partial f}{\partial \sigma_{kl}}\right). \tag{2.84}$$

From the above equation, the strain–stress relationship of elasto-plastic material can be expressed as

$$d\sigma_{ij} = \left(D^e_{ijkl} - \left(d\varepsilon_{kl} - d\lambda \frac{\partial f}{\partial \sigma_{kl}}\right) \frac{D^e_{ijmnl} \frac{\partial f}{\partial \sigma_{mn}} \frac{\partial f}{\partial \sigma_{pq}} D^e_{ijkl}}{\frac{\partial f}{\partial \sigma_{mn}} D^e_{mnpq} \frac{\partial f}{\partial \sigma_{pq}} - \frac{\partial f}{\partial L} \frac{\partial L}{\partial \varepsilon^p_{mn}} \frac{\partial f}{\partial \sigma_{mn}}}\right) d\varepsilon_{kl}. \tag{2.85}$$

We introduce the Drucker–Prager condition of the yield function, which is

$$f = q + \alpha p - \kappa, \tag{2.86}$$

where p and q are defined as follows:

$$p = \frac{1}{3}\sigma_{kk}, \tag{2.87}$$

$$q = \sqrt{\frac{3}{2} s_{ij} s_{ij}}, \tag{2.88}$$

where s_{ij} is deviatoric stress, defined as

$$s_{ij} = \sigma_{ij} - p\delta_{ij}. \tag{2.89}$$

Parameters α and k used in the yield function are dimensionless constants. Here, we choose α and k as follows:

$$\alpha = \frac{\tan \phi}{\sqrt{9 + 12 \cdot \tan^2 \phi}}, \tag{2.90}$$

$$\kappa = \frac{3 \cdot c}{\sqrt{9 + 12 \cdot \tan^2 \phi}}, \tag{2.91}$$

where c is cohesion and ϕ is an internal friction angle.

2.4.2.2 Jaumann Stress Rate

In the SPH method, deformation and motion of the spatial continuum are represented by motion of the discrete particles. A distinct feature of the Jaumann stress rate is that it can form a relationship between the stress rate and strain rate tensors.

$$\hat{\sigma}_{ij} = \dot{\sigma}_{ij} + \sigma_{ik}\omega_{kj} - \omega_{ik}\sigma_{kj}, \tag{2.92}$$

where w_{ij} is the rotation tensor and $\dot{\sigma}_{ij}$ is the Cauchy stress rate tensor. They are defined in the following:

$$\omega_{ij} = \frac{1}{2}\left(\frac{\partial v_i}{\partial x_j} - \frac{\partial v_j}{\partial x_i}\right), \tag{2.93}$$

$$\dot{\sigma}_{ij} = \frac{d\sigma_{ij}}{dt}. \tag{2.94}$$

The Jaumann stress rate $\hat{\sigma}_{ij}$ can be written as

$$\hat{\sigma}_{ij} = E_{ijkl}D_{kl}, \tag{2.95}$$

where E_{ijkl} is a fourth-order matrix that represents an elastic stiffness matrix. D_{ij} is a stretch tensor.

$$D_{ij} = \frac{1}{2}\left(\frac{\partial v_i}{\partial x_j} + \frac{\partial v_j}{\partial x_i}\right). \tag{2.96}$$

After the SPH approximations, D_{ij} can be rewritten in a discrete form:

$$D_{ij} = \frac{1}{2}\sum_j^N \frac{m_j}{\rho_j}\left[\left(v_x^j - v_x^i\right)\frac{\partial W_{ij}}{\partial x} + \left(v_y^j - v_y^i\right)\frac{\partial W_{ij}}{\partial y}\right]. \tag{2.97}$$

Similarly, we can obtain the discrete SPH form for Eq. (2.93) as

$$\omega_{ij} = \frac{1}{2}\sum_j^N \frac{m_j}{\rho_j}\left[\left(v_x^j - v_x^i\right)\frac{\partial W_{ij}}{\partial x} - \left(v_y^j - v_y^i\right)\frac{\partial W_{ij}}{\partial y}\right]. \tag{2.98}$$

The two-dimensional Jaumann stress rate, stretch tensor, and rotation tensor can be expressed as follows:

$$\dot{\sigma}_{xx} = \hat{\sigma}_{xx} + 2\omega_{xy}\sigma_{yx}, \tag{2.99}$$

$$\dot{\sigma}_{yy} = \hat{\sigma}_{yy} - 2\sigma_{xy}\omega_{xy}, \tag{2.100}$$

$$\dot{\sigma}_{xy} = \hat{\sigma}_{xy} - \sigma_{xx}\omega_{xy} + \sigma_{yy}\omega_{xy}, \tag{2.101}$$

$$D_{xx} = \frac{1}{2} \sum_{j}^{N} \frac{m_j}{\rho_j} \left[\left(v_x^j - v_x^i \right) \frac{\partial W_{ij}}{\partial x} \right],$$ (2.102)

$$D_{yy} = \frac{1}{2} \sum_{j}^{N} \frac{m_j}{\rho_j} \left[\left(v_y^j - v_y^i \right) \frac{\partial W_{ij}}{\partial y} \right],$$ (2.103)

$$D_{xy} = D_{yx} = \frac{1}{2} \sum_{j}^{N} \frac{m_j}{\rho_j} \left[\left(v_x^j - v_x^i \right) \frac{\partial W_{ij}}{\partial x} + \left(v_y^j - v_y^i \right) \frac{\partial W_{ij}}{\partial y} \right],$$ (2.104)

$$\omega_{xx} = \frac{1}{2} \sum_{j}^{N} \frac{m_j}{\rho_j} \left[\left(v_x^j - v_x^i \right) \frac{\partial W_{ij}}{\partial x} - \left(v_y^j - v_y^i \right) \frac{\partial W_{ij}}{\partial x} \right],$$ (2.105)

$$\omega_{yy} = \frac{1}{2} \sum_{j}^{N} \frac{m_j}{\rho_j} \left[\left(v_x^j - v_x^i \right) \frac{\partial W_{ij}}{\partial y} - \left(v_y^j - v_y^i \right) \frac{\partial W_{ij}}{\partial y} \right],$$ (2.106)

$$\omega_{xy} = -\omega_{xy} = \frac{1}{2} \sum_{j}^{N} \frac{m_j}{\rho_j} \left[\left(v_x^j - v_x^i \right) \frac{\partial W_{ij}}{\partial y} - \left(v_y^j - v_y^i \right) \frac{\partial W_{ij}}{\partial x} \right].$$ (2.107)

2.5   Numerical Aspects of the SPH Model

2.5.1   Initial Settings

Appropriate initial settings should be made for different computational problems. For specific requirements of varying computing cases, the initial state including the kernel function, type of particles and their coordinates, initial particle spacing, density, pressure, and velocity should be determined at the beginning of the calculation. h was chosen to be twice the particle spacing. The unit time step can be determined through

$$\Delta t = \min(0.4 \frac{h}{c}, \ 0.25 \sqrt{\frac{h}{f}}, \ 0.125 \frac{h^2}{v}),$$ (2.108)

where c is the speed of the sound, f is the external force, v is the viscosity coefficient, and 0.4, 0.25, and 0.125 are safety factors from experience.

2.5.2 Neighboring Particle Searching Algorithm

As mentioned above, SPH is a mesh-free method based on interactions with the closest neighboring particles. During the movement of each particle, the spatial location is constantly changing. Therefore, the neighbor list of each particle should be updated every time step. Therefore, creation of the neighbor list is important for high performance of the code. The efficiency of the SPH model strongly depends on the construction and use of this list.

There are currently two major approaches for determining the search area in the mesh-free methods: the Verlet neighbor list and linked-cell neighbor list methods. Put simply, the difference between these methods is that the Verlet neighbor list first defines a limit distance and then calculates the relative distance of each pair of particles. If this distance is less than the limit value, the particle can be considered a neighbor particle. The two-dimensional computational domain of each particle has a circular appearance. The linked-cell neighbor list generates a virtual mesh and determines the amount of computation by limiting the number of close grids. In this approach, the two-dimensional computational domain of each particle is a rectangle.

2.5.2.1 Linked-Cell Neighbor List Method

The linked-cell neighbor list method (Allen and Tildesley 1990; see Fig. 2.2) is a particle search method that takes searching scope determination as its principle. The greatest advantage is that its complexity is of the order O (N) rather than O (N^2). The entire calculation area is divided into several rectangular grids. Particles are allocated to different grids according to their coordinates. Side length r_{cell} is the

Fig. 2.2 Linked-cell neighbor list algorithm (based on Dominguez et al. 2011)

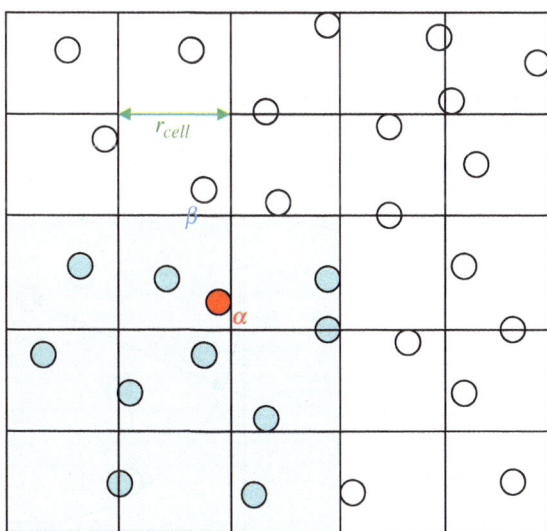

determinant of the mesh dimension, and its value directly impacts the effectiveness and accuracy of the method. After all the particles are included in the various grids, their motion information is stored in different lists and associated with each other. For example, in the calculation, if the pointer points to information of a particle in the list, the information of all the other particles in that list is obtained. For any problem, the complexity of this neighboring particle search algorithm is always of the order O (N). During the particle search, the calculation amount accumulates linearly with the change of particle number. However, large memory is required by this algorithm, and some cached data are lost.

2.5.2.2 Verlet Neighbor List Method

The Verlet neighbor list was named after Loup Verlet, who proposed this method in 1967 (Verlet 1967). The sketch of this algorithm is shown in Fig. 2.3. In its first use, the method greatly improved the computing speed for particle interactions in a small area. The core concept of the method is as follows. First, the distance between each particle pair is calculated, then a potential neighbor list is constructed in which particle pair distances are all within a "skin" layer radius r_v. In this list, only pairs of particles within cutoff radius r_c interact, resulting in another neighbor list. The neighbor list is updated every time step, whereas the potential neighbor list is updated every N time steps. The value of N was suggested to be 7 (Dominguez et al. 2011). According to Verlet's original paper (Verlet 1967), r_c and r_v can be taken as $2.5r$ and $3.2r$, respectively, where r is the radius of the interacting particles.

Figures 2.2 and 2.3 show the difference between these two methods when dividing the calculation area for the same problem. Each method has its strong points. The Verlet neighbor list method was selected for the numerical models in this monograph.

Fig. 2.3 Verlet neighbor list algorithm (reprinted from Dai et al. (2014), with permission of Elsevier)

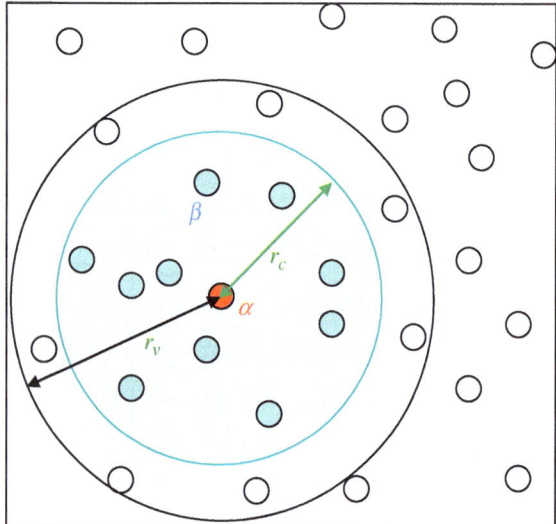

2.5.3 Kernel Function

As mentioned above, the kernel function is important in establishing the governing equations in the SPH method. The function determines the accuracy and validity of the approximation for the field function. The order, symmetry, and stability of the interpolation function should be considered when selecting the kernel function. The following lists some of the most frequently used kernel functions in the SPH literature.

A. *Gauss kernel*

$$W(x - x', h) = \left[\frac{1}{\sqrt{\pi}h}\right]^d \exp\left[-\frac{(x - x')^2}{h^2}\right], \qquad (2.109)$$

where d represents the dimension of space.

B. *Super Gauss kernel*

$$W(x - x', h) = \left[\frac{1}{\sqrt{\pi}h}\right]^d \left[\frac{5}{2} - \frac{(x - x')^2}{h^2}\right] \exp\left[-\frac{(x - x')^2}{h^2}\right]. \qquad (2.110)$$

C. *Exponential kernel*

$$W(x - x', h) = \frac{1}{8\pi h^3} \exp\left[-\frac{(x - x')}{h}\right]. \qquad (2.111)$$

D. *Cubic spline function*

$$W(s, h) = \frac{C}{h^d} \begin{cases} 1 - \frac{3}{2}s^2 + \frac{3}{4}s^3 & 0 \le s \le 1 \\ \frac{1}{4}(2 - s)^2 & 1 \le s \le 2 \\ 0 & s \ge 2. \end{cases} \qquad (2.112)$$

Here, $s = |x - x'|/h$ and C is 2/3, $10/7\pi$, and $1/\pi$, respectively, in one-, two-, and three- dimensional space, for the unity requirement.

E. *B-spline function*

Monaghan and Lattanzio (1985) devised the following smoothing function based on the cubic spline functions known as the B-spline function (Fig. 2.4):

$$W(s, h) = a_d \begin{cases} \frac{2}{3} - s^2 + \frac{1}{2}s^3 & 0 \le s \le 1 \\ \frac{1}{6}(2 - s)^3 & 1 \le s \le 2 \\ 0 & s \ge 2. \end{cases} \qquad (2.113)$$

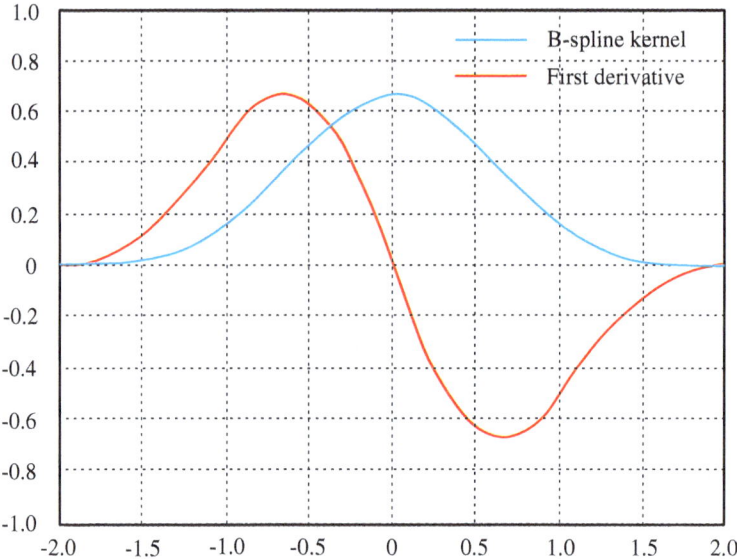

Fig. 2.4 B-spline kernel and its first derivative

Here, s is the same as above. In one-, two-, and three-dimensional space, $a_d = 1/h$, $15/7 \ \pi h^2$, and $3/2 \ \pi h^2$, respectively. The cubic spline function has been the most widely used smoothing function in the SPH literature, since it resembles a Gaussian function while having a narrower compact support.

In this monograph, the B-spline function was selected as the kernel function. Taking the derivative of the above equation,

$$\nabla_x \omega = \frac{\partial \omega}{\partial s} \cdot \frac{\partial s}{\partial r} \cdot \frac{\partial r}{\partial x} = \alpha_d \times \begin{cases} \left(-2s + \frac{3}{2}s^2\right)\left(\frac{1}{h}\right)\left(\frac{x-x'}{r}\right) & 0 \leq s < 1 \\ -\frac{1}{2}(2-s)^2\left(\frac{1}{h}\right)\left(\frac{x-x'}{r}\right) & 1 \leq s < 2 \\ 0 & s \geq 2. \end{cases}$$

$$(2.114)$$

2.5.4 Free-Surface Boundary Treatment

Boundary conditions vary with the research subject in this monograph. In the SPH fluid dynamics mode, it is important to distinguish the free-surface boundaries. If the particle density meets the following requirement, then the particle is considered a free-surface particle:

$$\rho < \beta \rho_0. \tag{2.115}$$

where ρ_0 is the real density of the fluid and β is a parameter less than 1, with a value between 0.8 and 0.98. The specific value can be determined by experiment.

In the SPH solid mechanics mode, various distinguishing marks of the particles in different collections are tagged to determine whether a particle is involved in the calculation and, if so, its degrees of freedom. For example, N is set as the distinguishing mark. If $N = 1$, the particle is fixed in the X direction; if $N = 2$, it is fixed in the Y direction; if $N = 3$, it is fixed in both the X and Y directions.

2.5.5 Averaging the Velocity Field

In a Lagrangian method, the new position of a particle is derived by time integration of the velocity at every moment. A variant called XSPH was proposed in the SPH literature, with the goal of modifying and smoothing SPH particle movement. However, the acceleration equation remained unchanged. The smoothed velocity was defined by an average over the velocities of neighboring particles, according to Monaghan (1992):

$$v_i = v_i + \chi \sum_{j=1}^{N} \frac{m_j}{\rho_{ij}} \left(v_j - v_i \right) W_{ij}, \tag{2.116}$$

where χ is a constant, the value of which is between 0 and 1. Numerical experiments show that $\chi = 0.5$ is effective at smoothing local fluctuations of velocity (Monaghan 2002). ρ_{ij} denotes an average density of particles i and j.

2.5.6 Solid Boundary Treatment

Near the boundary, only neighboring particles inside the boundary contribute to the SPH summation of particle interaction; no contribution comes from outside the boundary because there are no particles there (Fig. 2.5). Therefore, SPH simulations may have a problem of particle deficiency near the boundary, which leads to inaccurate solutions.

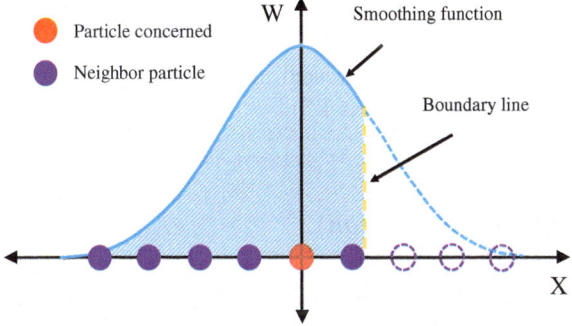

Fig. 2.5 SPH particle approximations for the one-dimensional case

To address this problem, some improvements have been proposed to deal with the boundary condition. Among such methods, free-slip boundaries have been widely used in SPH simulation for free-surface flows. This type of simulation uses boundary particles that exert strong repulsive forces that prevent SPH particles from penetrating the solid surface (Monaghan 1994). These boundary particles do not contribute to the density of the free SPH particles. Libersky et al. (1993) introduced ghost particles of opposite velocity to reflect a symmetrical surface boundary condition. Later, a more general treatment was proposed (Randles and Libersky 1996), in which all the ghost particles were assigned the same boundary field variable to calculate the values of the interior particles. The present work incorporates a no-slip boundary condition first proposed by Morris et al. (1997), which is similar to that used by Takeda et al. (1994). Figure 2.5 illustrates the principal concept for a curved boundary. The yellow dashed line represents the boundary line. The red particle is the flow particle concerned. Pseudo-boundary particles (purple) are set on and outside the boundary line. They contribute to the density of the flow particles, thus avoiding the problem of particle deficiency near the solid boundary. A tangent plane to the boundary surface is defined and normal distances of the fluid particle and boundary particle to this tangent plane, d_A and d_B, are calculated. The velocity of particle A is extrapolated across the tangent plane, assuming zero velocity on the plane itself (no-slip). Then, the velocity of each image boundary particle can be calculated using

$$V_B = -(d_B/d_A) \cdot V_A, \tag{2.117}$$

The artificial velocities are used here to calculate the viscous force rather than to update boundary particle positions. The relative velocities between fluid and boundary particles V_{AB} can then be defined by

$$V_{AB} = V_A - V_B = (1 + d_B/d_A) \cdot V_A. \tag{2.118}$$

According to Morris et al. (1997), viscous forces from the solid boundary can be calculated using relative velocities. To avoid exerting very large viscous forces when the fluid particles approach the boundary surface too closely, the magnitude of V_{AB} must be restricted through the following equations:

$$V_{AB} = \beta \cdot V_A, \tag{2.119}$$

$$\beta = \min(\beta_{\max}, \, 1 + d_B/d_A), \tag{2.120}$$

where β is a safety parameter to prevent numerical singularities and $\beta_{\max}$ is an empirical parameter (often chosen as 1.5).

2.5.7 Acceleration Calculation

Acceleration of a particle can be generated by internal and external forces. In the fluid model, particle density should initially be calculated using the continuity equation. Second, free-surface particles should be caught and their densities modified. Then, the pressure is calculated via the equation of state, and acceleration is caused by the external force. The final acceleration can then be obtained through the momentum equation. For the solid model, the Jaumann stress rate is used to calculate the change of stress and strain of the particles. Then, the final acceleration is determined by accumulating those from internal and external forces.

2.5.8 Integration Schemes

To calculate forces acting on the SPH particles after discretization, a neighbor list is created for each particle in the current configuration. In the subsequent calculations, the neighbor list is referred to, so the total force acting on a particle can be determined. Since the SPH method reduces the original PDEs to sets of ODEs, any stable time-step algorithm for ODEs can be introduced. The Velocity Verlet algorithm (Ercolessi 1997), similar to the leapfrog method, is introduced here for the time integration to update positions, velocities, and accelerations:

$$x(t + \Delta t) = x(t) + v(t)\Delta t + \frac{1}{2}a(t)\Delta t^2, \tag{2.121}$$

$$v(t + \Delta t/2) = v(t) + \frac{1}{2}a(t)\Delta t, \tag{2.122}$$

$$v(t + \Delta t) = v(t + \Delta t/2) + \frac{1}{2}a(t + \Delta t)\Delta t, \tag{2.123}$$

$$a(t) = \frac{1}{\rho}f(t). \tag{2.124}$$

2.5.9 Outputs of Calculation

Given the Lagrangian property of the SPH method, information on each particle, such as density, position, velocity, and pressure, can be captured at every time step. According to the actual needs of a study, these variable values can be selectively output, easily visualized, and rapidly analyzed, and time history curves can be readily obtained.

2.6 Summary

SPH is a novel mesh-free particle method based on a pure Lagrangian description. The basic idea of the method is that a continuous fluid is represented by a set of arbitrarily distributed particles. By providing accurate and stable numerical solutions for hydrodynamic equations and tracking the movements of each particle, the mechanical behavior of the full system can be determined. SPH is characterized by a mesh-free, adaptive, and Lagrangian description, which makes it suitable for handling the problems of large deformation and free surfaces.

On a theoretical level, SPH can accurately describe the mechanical process as long as there are sufficient numbers of particles. Although the precision of results depends on particle arrangement, the requirement for this arrangement is much less stringent than the demands of a grid. Without the grid, there is no connectivity between the particles. Therefore, severe mesh distortions caused by large deformation are avoided, thereby improving computational accuracy. In addition, SPH can conveniently handle the interface of different materials. Another advantage of the method is the Lagrangian description, which avoids the difficult interface of the grid and material. Hence, the SPH method is especially suitable for solving complex flow problems. In particular, the method has the following advantages: (1) there is no migration term in the PDEs, so the program design is simple and efficient; (2) it is easy to track the time history of all field variables for all particles; (3) it can automatically exert the boundary condition to track the free surfaces, material interfaces, and moving boundaries; (4) it is easy to handle irregular and complex geometry shapes. SPH is therefore a novel and promising method for computational fluid mechanics.

In summary, we recap the following major points:

(1)   We briefly summarized the origin of the SPH method and its main concepts. Two core approximations, those of the kernel and particle, were described in detail.

(2)   Based on SPH basic theories, SPH formulations for the N–S equations in a Lagrangian description were established. It was pointed out that accuracy near both free boundaries and material interfaces is the criterion for elevating the approximate formats.

(3)   The key issue for establishing the SPH formulations for elasto-plastic mechanics is analysis of the constitutive relationship between material stress and strain, and internal particle motion. Based on the constitutive model in the form of the Jaumann stress rate, the SPH model can be applied to many types of materials.

(4)   The implicit algorithm of Poisson's equation for calculating the pressure in SPH can improve the stability and accuracy of the numerical solution in the fluid dynamics model.

(5)   Computational efficiency and accuracy are key elements for evaluating a numerical simulation method. Relative to the mesh-based method, the unique techniques in the SPH method, including the special treatment in the solution method, efficient particle searching, and selection of the smoothing kernel function, help SPH analyze physical problems efficiently and accurately.

References

Allen, M. P., & Tildesley, D. J. (1990). *Computer simulation of liquids*. New York: Oxford University Press.

Chen, J. K., Beraun, J. E., & Jih, C. J. (1999). An improvement for tensile instability in smoothed particle hydrodynamics. *Computational Mechanics, 23*, 279–287.

Cleary, P. W., & Monaghan, J. J. (1999). Conduction modeling using smoothed particle hydrodynamics. *Journal of Computational Physics, 148*, 227–264.

Dai, Z. L., Huang, Y., Cheng, H. L., & Xu, Q. (2014). 3D numerical modeling using smoothed particle hydrodynamics of flow-like landslide propagation triggered by the 2008 Wenchuan earthquake. *Engineering Geology,*. doi:10.1016/j.enggeo.2014.03.018.

Dominguez, J. M., Crespo, A. J. C., Gomez-Gesteira, M., & Marongiu, J. C. (2011). Neighbour lists in smoothed particle hydrodynamics. *International Journal for Numerical Methods in Fluids, 67*(12), 2026–2042.

Dyka, C. T. (1994). *Addressing tension instability in SPH methods.* Technical Report NRL/MR/6384, NRL.

Ercolessi, F. (1997). *A molecular dynamics primer* (pp. 24–25). Trieste: Spring College in Computational Physics, ICTP.

Evard, A. E. (1988). Beyond N-Body: 3D cosmological gas dynamics. *Monthly Notices of the Royal Astronomical Society, 235*, 911–934.

Gingold, R. A., & Monaghan, J. J. (1977). Smoothed particle hydrodynamics: Theory and application to non-spherical stars. *Monthly Notices of the Royal Astronomical, 181*, 375–389.

Huang, Y., Zhang, W. J., Mao, W. W., & Jin, C. (2011). Flow analysis of liquefied soils based on smoothed particle hydrodynamics. *Natural Hazards, 59*(3), 1547–1560.

Johnson, G. R., & Beissel, S. R. (1996). Normalized smoothed functions for SPH impact computations. *International Journal for Numerical Methods in Engineering, 39*(16), 2725–2741.

Libersky, L. D., Petschek, A. G., & Carney, T. C. (1993). High strain Lagrangian hydrodynamics: A three dimensional SPH code for dynamic material response. *Journal of Computational Physics, 109*(1), 67–75.

Liu, G. R., & Liu, M. B. (2003). *Smoothed particle hydrodynamics-a meshfree particle method.* New Jersey: World Scientific Publishing Company.

Lucy, L. B. (1977). A numerical approach to the fission hypothesis. *The Astronomical Journal, 82*(12), 1013–1024.

Monaghan, J. J. (1988). An introduction to SPH. *Computer Physics Communication, 48*, 89–96.

Monaghan, J. J. (1992). Smoothed particle hydrodynamics. *Annual Review of Astronomy and Astrophysics, 30*, 543–574.

Monaghan, J. J. (1994). Simulating free surface flows with SPH. *Journal of Computational Physics, 110*, 399–406.

Monaghan, J. J. (2002). SPH compressible turbulence. *Monthly Notices of the Royal Astronomical Society, 335*(3), 843–852.

Monaghan, J. J., & Gingold, R. A. (1983). Shock simulation by the particle method SPH. *Journal of Computational Physics, 52*, 374–389.

Monaghan, J. J., & Lattanzio, J. C. (1985). A refined particle method for astrophysical problems. *Astronomy & Astrophysics, 149*(1), 135–143.

Morris, J. P., Fox, P. J., & Zhu, Y. (1997). Modeling low Reynolds number incompressible flows using SPH. *Journal of Computational Physics, 136*, 214–226.

Moriguchi, S. (2005). CIP-based numerical analysis for large deformation of geomaterials. *Ph.D. Dissertation,* Gifu University, Japan.

Nonoyama, H. (2011). Numerical application of SPH Method for deformation, failure and flow problems of geomaterials. *Ph.D. thesis,* Gifu University, Japan.

Randles, P. W., & Libersky, L. D. (1996). Smoothed particle hydrodynamics: Some recent improvements and applications. *Computer Methods in Applied Mechanics and Engineering, 139*, 375–408.

Swegle, J. W., Hicks, D. L., & Attaway, S. W. (1995). Smoothed particle hydrodynamics stability analysis. *Journal of Computational Physics, 116*, 123–134.

Takeda, H., Miyama, S. M., & Sekiya, M. (1994). Numerical simulation of viscous flow by smoothed particle hydrodynamics. *Progress of Theoretical Physics, 92*(5), 939–960.

Verlet, L. (1967). Computer "Experiments" on classical fluids. I. Thermodynamical properties of Lennard–Jones molecules. *Physical Review, 159*(1), 98–103.

Chapter 3
Computer Procedure and Visualization Software

In the last chapter, the governing equations were discretized into Smoothed Particle Hydrodynamics (SPH) form, and constitutive laws in hydrodynamics and elasto-plastic mechanics were incorporated. In this chapter, the corresponding computer procedure, which is applicable for geo-disaster analysis and modeling, is developed in the FORTRAN language environment. The function of each module in the procedure of this monograph is detailed with a calculation flowchart. Based on the calculation procedure, visual simulation software is developed with a Windows interface for the geo-disaster analysis and modeling, resulting in a greatly improved computational efficiency for the SPH method.

3.1 Flowchart of the SPH Procedure

SPH procedures based on hydrodynamics and elasto-plastic mechanics were developed according to the introduction and summary above.

The computer procedure developed in this monograph can be applied to simulate large deformation of soil material in the framework of two different theories of mechanics. The main functions of the procedure are as follows.

1. The following are the types of problems that can be solved by the hydrodynamics SPH procedure:

   a. The freedom of motion of water mass ($NNB = 1$).
   b. Boil flow based on the Bingham fluid constitutive model ($NNB = 2$).

2. The following are the types of problems that can be solved by the elasto-plastic SPH procedure:

   a. Deformation problem of perfectly elastic material ($EP = 1$).
   b. Deformation problem of soil based on the Drucker-Prager model ($EP = 2$).
   c. Deformation problem of soil based on the Modified Cam-Clay model ($EP = 3$).

© Springer-Verlag Berlin Heidelberg 2014
Y. Huang et al., *Geo-disaster Modeling and Analysis: An SPH-based Approach*,
Springer Natural Hazards, DOI 10.1007/978-3-662-44211-1_3

3.1.1 Hydrodynamics SPH Program

The main flowchart of the hydrodynamics SPH program is shown in Fig. 3.1.
The following describes the main function of each computing module.
Module 1: Data input
Function:

1. Input the identification mark of the computational problem (NNB = 1, the freedom movements of water mass; NNB = 2, soil flow based on the Bingham fluid constitutive model) and parameters including density, velocity, external force, and others.
2. Computer memory space allocation. To save computer memory, dynamic arrays are used to store variables.

   Module 2: Initial settings
   Function:

1. Calculate particle mass and external force and set computation parameters.
2. Calculate initial density and set the initial free boundary. See below flowchart (Fig. 3.2).

Fig. 3.1 Main flowchart
of the hydrodynamics SPH
program

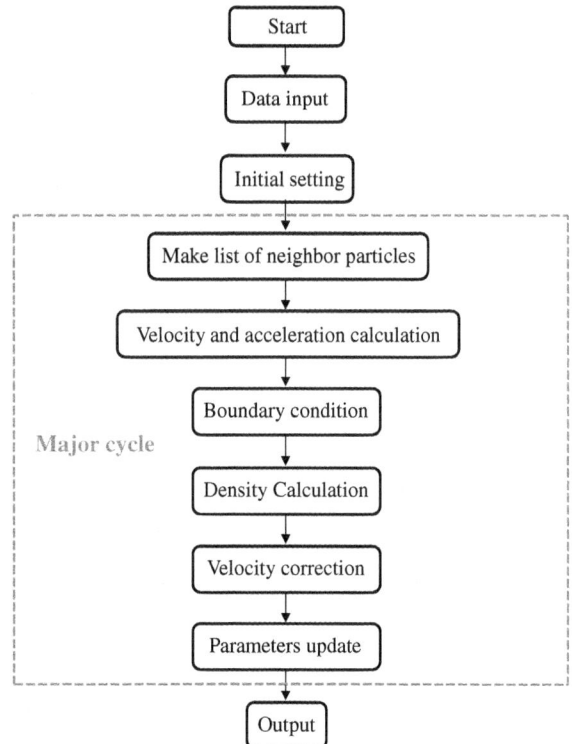

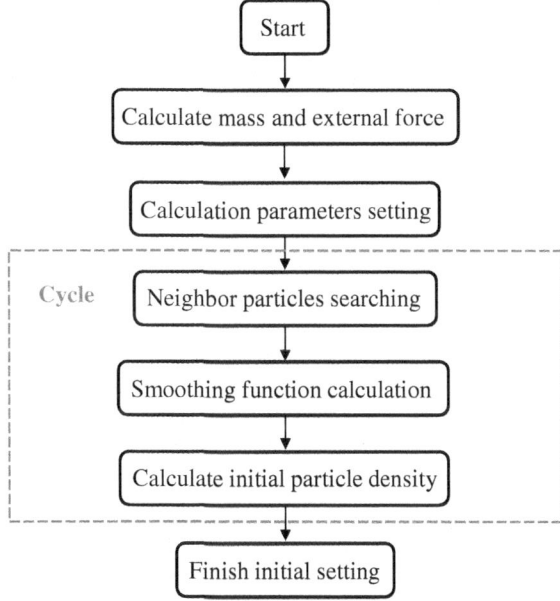

Fig. 3.2 Flowchart of the initial setting module

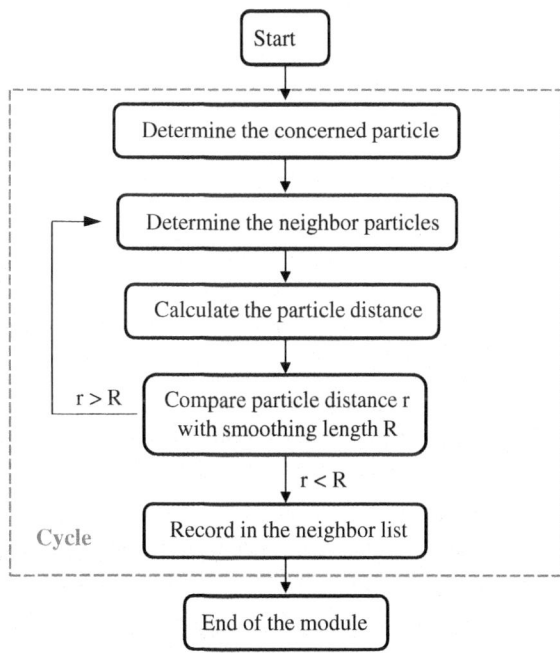

Fig. 3.3 Flowchart of particle search module

Module 3: Neighboring particle search

Function: Determine neighboring particles within the influence radius and establish the neighbor list. See below flowchart (Fig. 3.3).

Fig. 3.4 Flowchart of
calculation module of
velocity, acceleration, and
position

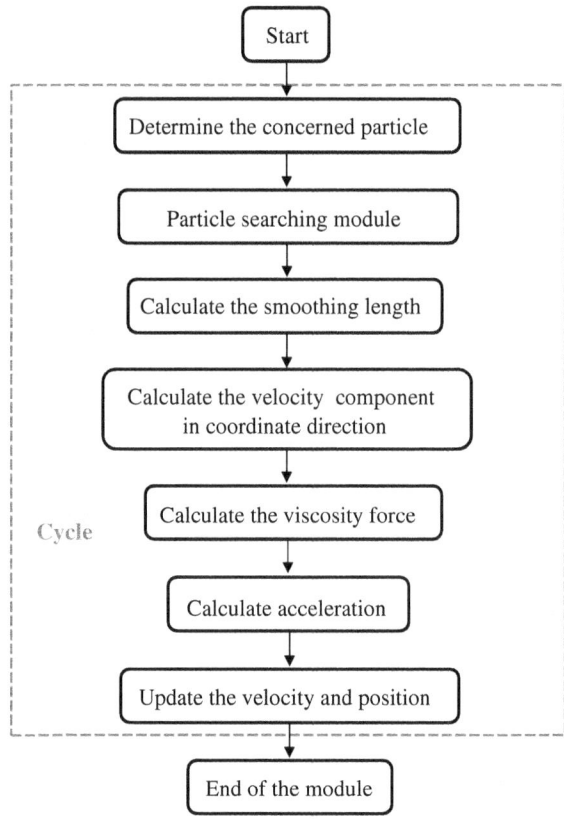

Module 4: Calculation module of velocity, acceleration, and position
Function: Calculate the effect on particle velocity, acceleration, and position
from the viscous force caused by the motion of neighboring particles. See below
flowchart (Fig. 3.4).

Module 5: Boundary setting
Function: Setting the condition of calculation, including degrees of freedom of
the boundary particles and moving particles. See below flowchart (Fig. 3.5).

Module 6: Density calculation
Function: Calculate particle density and identify particles on the free surface at
every time step. See below flowchart (Fig. 3.6).

Module 7: Velocity correction
Function: Calculate the difference in pressure caused by changes in the number
of neighboring particles in its supporting region. Then, calculate the effect on the
particle acceleration and correct the velocity. See below flowchart (Fig. 3.7).

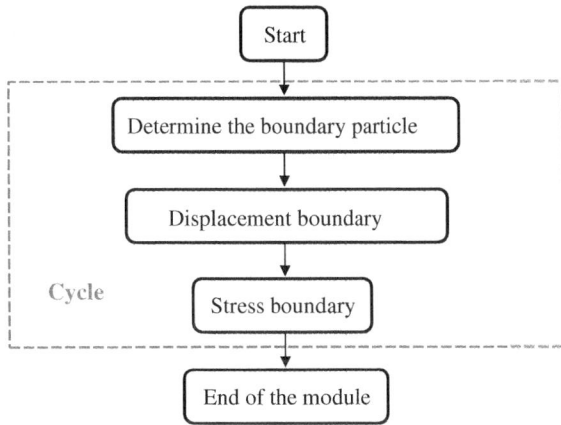

Fig. 3.5 Flowchart of boundary setting module

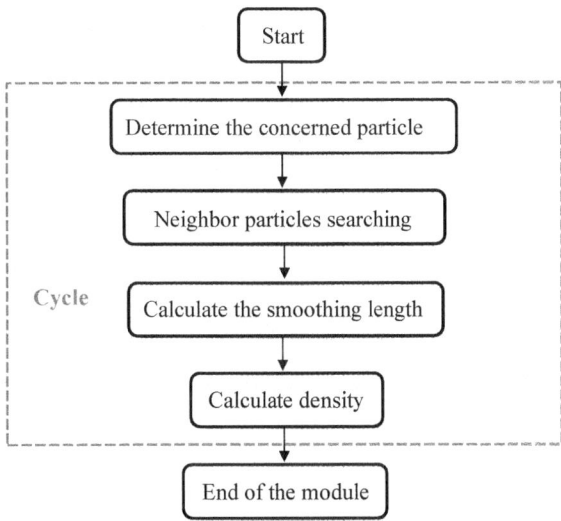

Fig. 3.6 Flowchart of density calculation module

Module 8: Information update
Function:

1. Calculate coordinates, velocity, and acceleration of each particle after each calculation cycle.
2. Update coordinates and velocities as the initial condition of the next step. See
   below flowchart (Fig. 3.8).

   Module 9: Data output
   Function: Output all quantities of each particle. See below flowchart (Fig. 3.9).

Fig. 3.7 Flowchart of
velocity correction module

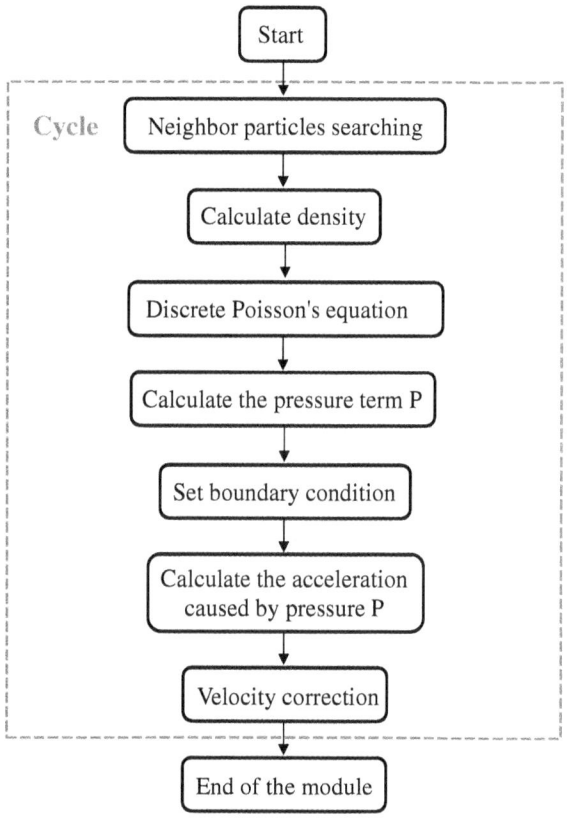

Fig. 3.8 Flowchart of
information update module

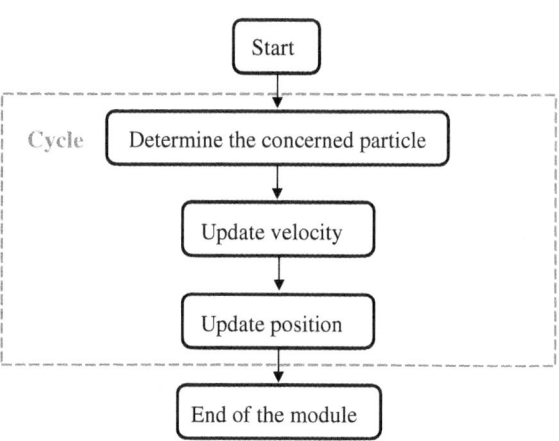

Fig. 3.9 Flowchart of data output module

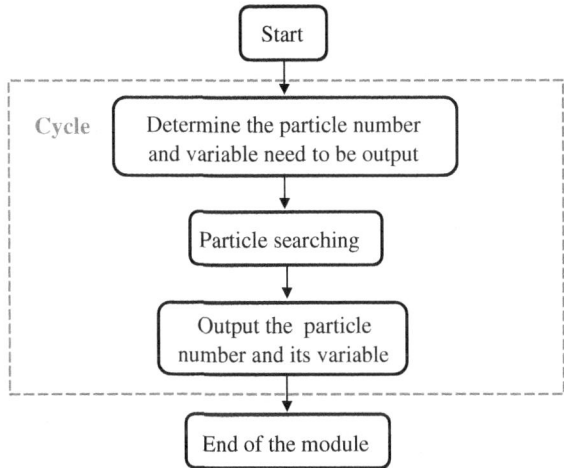

3.1.2 Elasto-Plastic SPH Program

The main flowchart of the elasto-plastic SPH program is shown in Fig. 3.10.
   Module 1: Data input
   Function:

1. Input the identification mark of the computational problem (EP = 1, deformation problem of perfectly elastic material; EP = 2, deformation problem of soil based on Drucker-Prager model; EP = 3, deformation problem of soil based on Modified Cam-Clay model) and parameters including density, velocity, external force, and others.
2. Computer memory space allocation.

   Module 2: Initial setting
   Function:

1. Calculate the initial value of particle quantities and set the computation parameters.
2. Set the initial stress state of the particles. See below flowchart (Fig. 3.11).

   Module 3: Neighboring particle searching
   Function: The same as that in the hydrodynamics SPH program. The only difference is the choice of smoothing length. See below flowchart (Fig. 3.12).

   Module 4: Stress and strain calculation
   Function: Calculate strain and stress of the particles. See below flowchart (Fig. 3.13).

   Module 5: Boundary setting
   Function: Setting the condition of calculation, including degrees of freedom of the boundary particles and moving particles. See below flowchart (Fig. 3.14).

Fig. 3.10 Main flowchart
of the elasto-plastic SPH
program

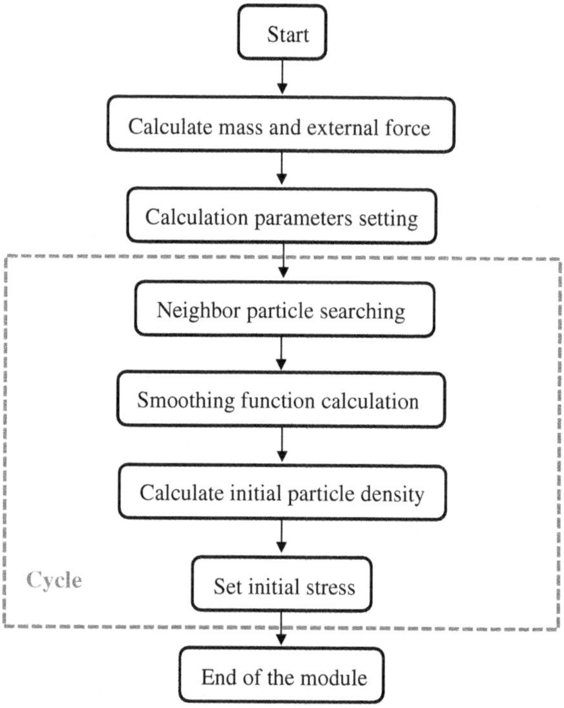

Fig. 3.11 Flowchart of
initial setting module

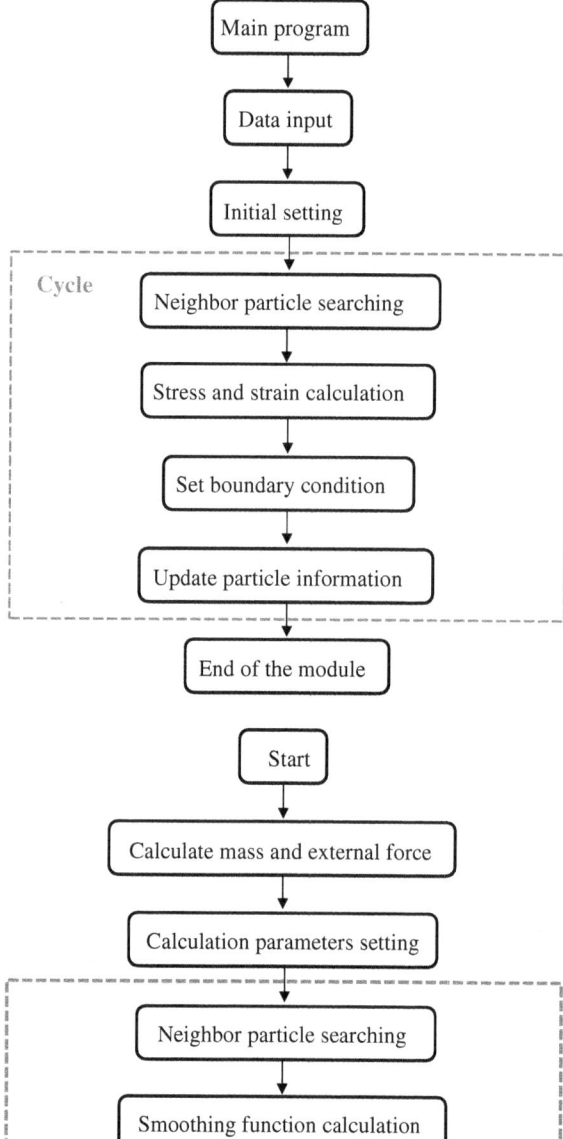

Fig. 3.12 Flowchart of
particle search module

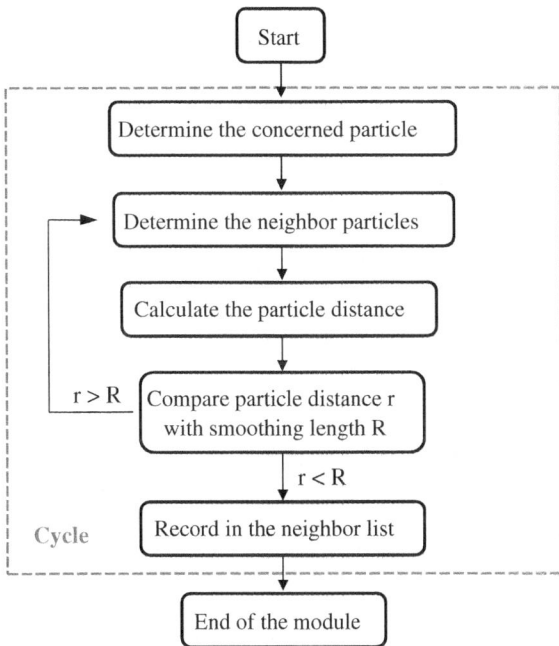

Module 6: Information update

Function: Update coordinates and velocity of each particle after each calculation cycle. These values are used as the initial condition of the next step. See
below flowchart (Fig. 3.15).

Module 7: Data output

Function: Output all quantities of each particle. See below flowchart (Fig. 3.16).

The SPH program can be developed following the steps above. Code for the
stress and strain calculation module is in Appendix A.

3.2 Visualization Software Development

In the last section, the SPH program based on hydrodynamics and elasto-plastic
mechanics was developed for large deformation simulation of soil material.
However, traditional numerical simulation is both time-consuming and prone to
error when constructing a geometric model, writing complicated computer programs, and analyzing complex computation results. With the rapid development of
computer technology, visualization techniques in scientific research have become
more important. Visualization can transform complex computation results into
vivid graphics, improve user understanding of complicated and highly interrelated
information, avoid substantial duplication of work, and boost the efficiency of

Fig. 3.13 Flowchart of stress
and strain calculation module

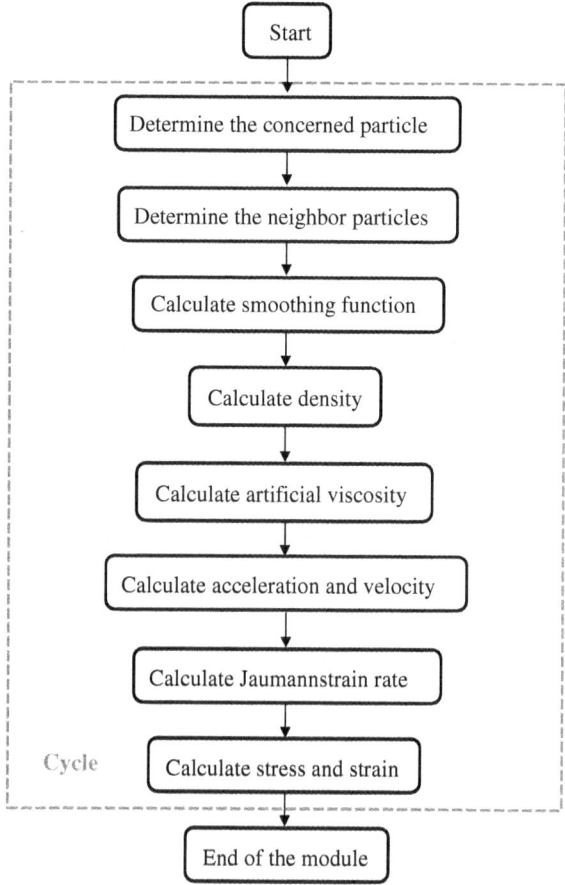

Fig. 3.14 Flowchart of
boundary setting module

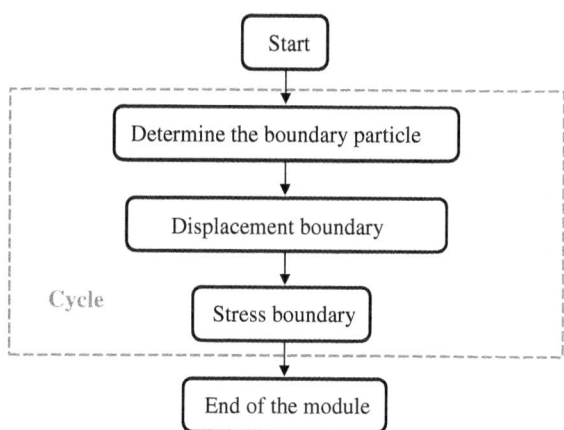

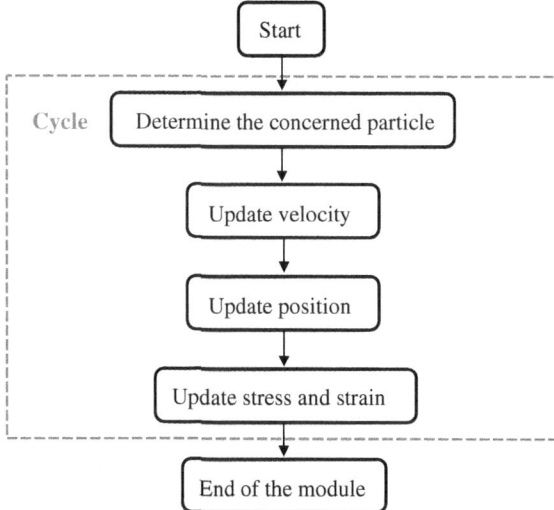

Fig. 3.15 Flowchart of information update module

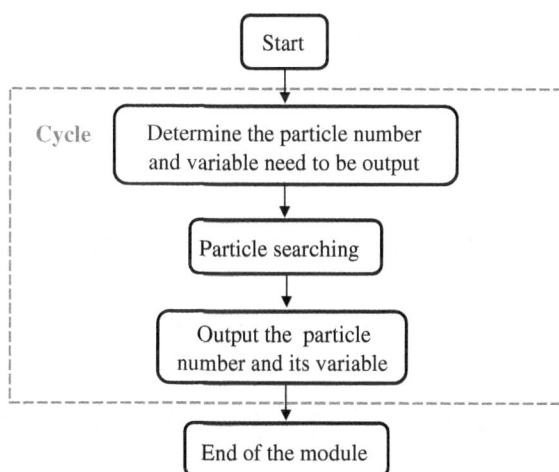

Fig. 3.16 Flowchart of data output module

scientific research. Therefore, visualization in the simulation of large deformation of soil material has become a major focus.

In this section, visualization software is developed to simulate the fluidized movement of soil material. This software is called Visual SPH and is based on the SPH method. The software has a user-friendly interface to facilitate human–machine communication and enhance simulation efficiency. It aims to provide a clear user perspective of the full range of motion of soil material and to transform large amounts of abstract data into plots of velocity, displacement, and impact force versus time. This helps the user understand the motion of soil material and obtain its dynamic behavior.

3.2.1 Programming Language

Visual SPH has been implemented successfully in Visual Basic (VB) 6.0, supported by the FORTRAN language and Matrix VB plug-ins.

Of the various programming languages available, VB is a natural choice, since it is efficient, powerful, and easy to learn. Supported by Windows Application Program Interface (API) functions, the Dynamic Link Library (DLL), and Object Linking and Embedding (OLE), VB can be efficiently used to implement powerful application systems software with rich graphical interfaces in a Windows environment.

FORTRAN is an acronym for "Formula Translation" and is a powerful tool for scientific computing. There are two main calculation subroutines in Visual SPH, fc.exe and dlsph.exe. These calculate the velocity and displacement increment for each particle during the flow, respectively. The subroutines are written in FORTRAN and called through "WinExec" functions in the Windows API.

Matrix VB plug-ins is a programming interface for VB provided by MathWorks. The dynamic link file *MMatrix.dll* internalizes Matrix VB into VB, allowing the use of all functions in the MATLAB computing environment to compute large matrices and draw graphs directly.

This kind of mixed programming approach easily combines the powerful computing capabilities of FORTRAN, efficient visualization functions of VB, and drawing functions of Matrix VB. All these complement each other, thereby enhancing the quality and efficiency of the software.

3.2.2 Program Description

(1) Program structure and window style

The structure of Visual SPH is shown in Fig. 3.17. "FrmMain" is the main window and controls the program interface. Its primary role is to call and communicate with each sub-form. A menu bar is provided with five main menus, "File," "Data Input," "Compute," "Result," and "Help," each of which contains several submenus that can be used to establish models, input parameters, and compute and output results.

The program uses interactive user-machine interfaces to provide convenient communication between user and computer and adopts the Multiple Document Interface (MDI), in which all windows can be moved freely to any position on the screen. Therefore, users can open multiple windows simultaneously to view or compare data and documents.

(2) Preprocessing

Preprocessing involves three steps: model setup, parameter input, and discretization of the problem domain.

The "Model" window, which is divided into two areas, is used to set up a geometric landslide model (Fig. 3.18). The left side of the window shows a black line

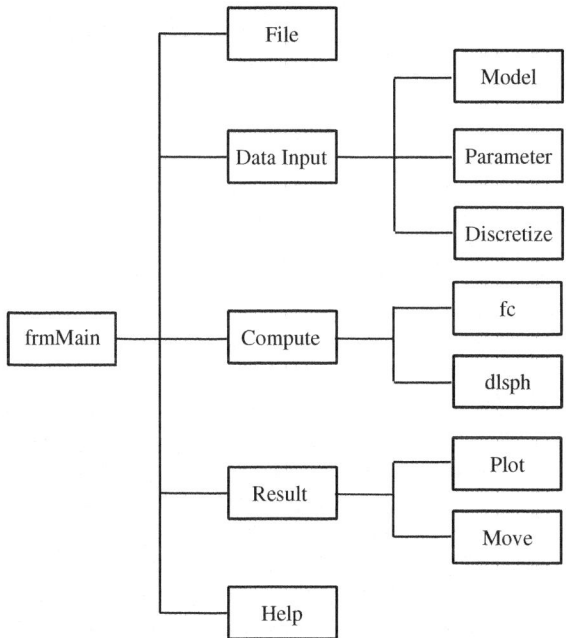

Fig. 3.17 Program structure (reprinted from Huang et al. (2011), with permission from Springer)

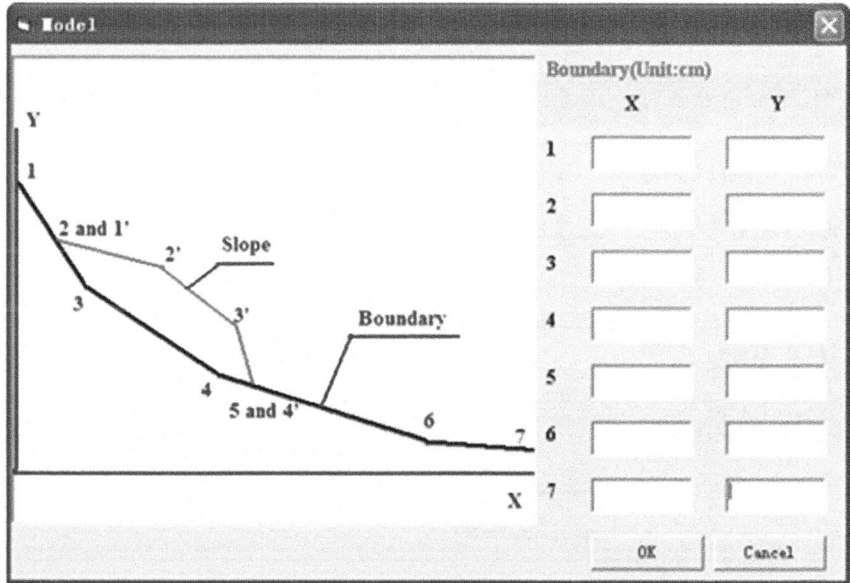

Fig. 3.18 Model window (reprinted from Huang et al. (2011), with permission from Springer)

denoting the boundary with several control points, and a red line representing the fluid contour with four control points. These two key lines identify the configuration of the landslide and trajectory. The right side of the window has a set of text-boxes, allowing users to input coordinate values of corresponding control points. The program calls the "Plot" function from the Matrix VB plug-ins to connect all control points and draw the model structure. Therefore, users can easily establish a geometric model simply by inputting the coordinate values, thereby avoiding tedious drawing steps and saving time.

Model parameters are input via the "Parameters" window, according to certain conditions. The interface contains eight dialog boxes, providing eight input points for parameters (Fig. 3.19). The first four parameters are physical properties of the fluid, i.e., density, viscosity, and horizontal and vertical acceleration. The remaining four are total steps, time interval, and intervals for output to the data file and screen. Corresponding text labels and the units for the parameters are alongside the dialog boxes. Input values are assigned directly to variables used in calculations.

The "Discretize" submenu is mainly used to discretize the problem domain into a set of particles carrying field variables. The program first defines equations of the line segments of the boundary and fluid contour through coordinate values of the control points, and then uses a do-loop cycle to discretize the model from left to right and top to bottom along the line segments. Boundary and fluid particles are distinguishable by values of the variables "pmat" and "pnbf." Thereafter, the program outputs two documents, "1.dat," which contains the coordinate values of each particle and its acceleration, and "2.dat," which records the total number of particles, density, viscosity, total steps, and other information. Finally, the discretized model is plotted as in Fig. 3.20.

Fig. 3.19 Parameter input window (reprinted from Huang et al. (2011), with permission from Springer)

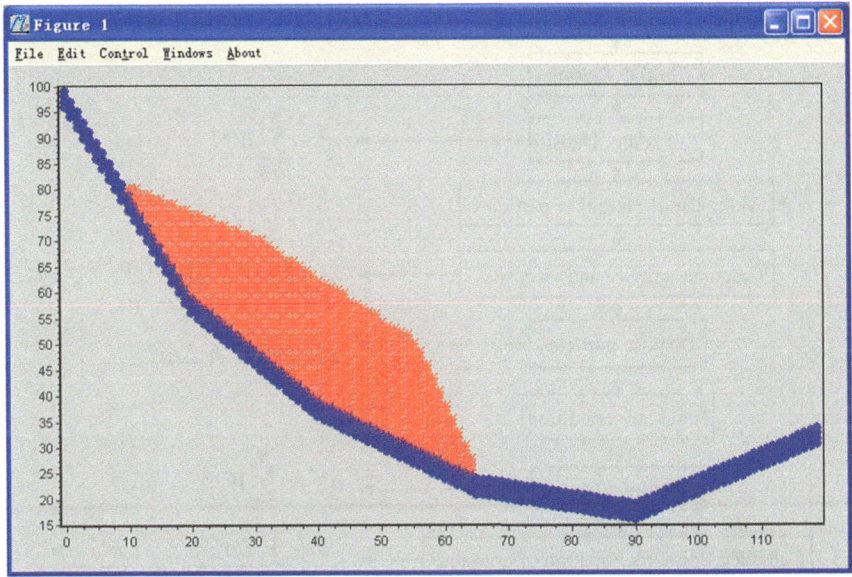

Fig. 3.20 Discretized model (reprinted from Huang et al. (2011), with permission from Springer)

(3) Computation

The "Compute" menu uses the Windows API function "WinExec" to call subroutines "fc.exe" and "dlsph.exe," written in FORTRAN. Subroutine fc.exe reads the particles' initial coordinate values and their physical parameters, whereas dlsph.exe calculates the velocity and displacement increment for all particles at each time step. The subroutine flowchart is shown in Fig. 3.21. Given the Lagrangian nature of the SPH method, after the calculations, physical parameters of all particles, such as the location vector, density, velocity, and pressure, can be obtained at each step. Owing to the need for further research, the program outputs the particles' dynamic behavior including velocity, displacement, and pressure to the file STEP.DAT. This file is used to generate the parameters' time curves and the particles' coordinate values to the file MOVE.MGF, which is used to generate animation and simulate landslide fluidization movement.

(4) Post-processing

Post-processing is essential in visual simulation. Its primary aim is to turn abstract computing results into static charts and dynamic graphics that can be easily understood. Post-processing is done mainly through the "Result" menu, which contains two submenus, "Plot" and "Move."

The "Plot" window mainly shows the fundamental dynamic behavior of particles, such as velocity, displacement, and pressure, which change over time during the flow process. The window is separated into two areas, with a picture frame at

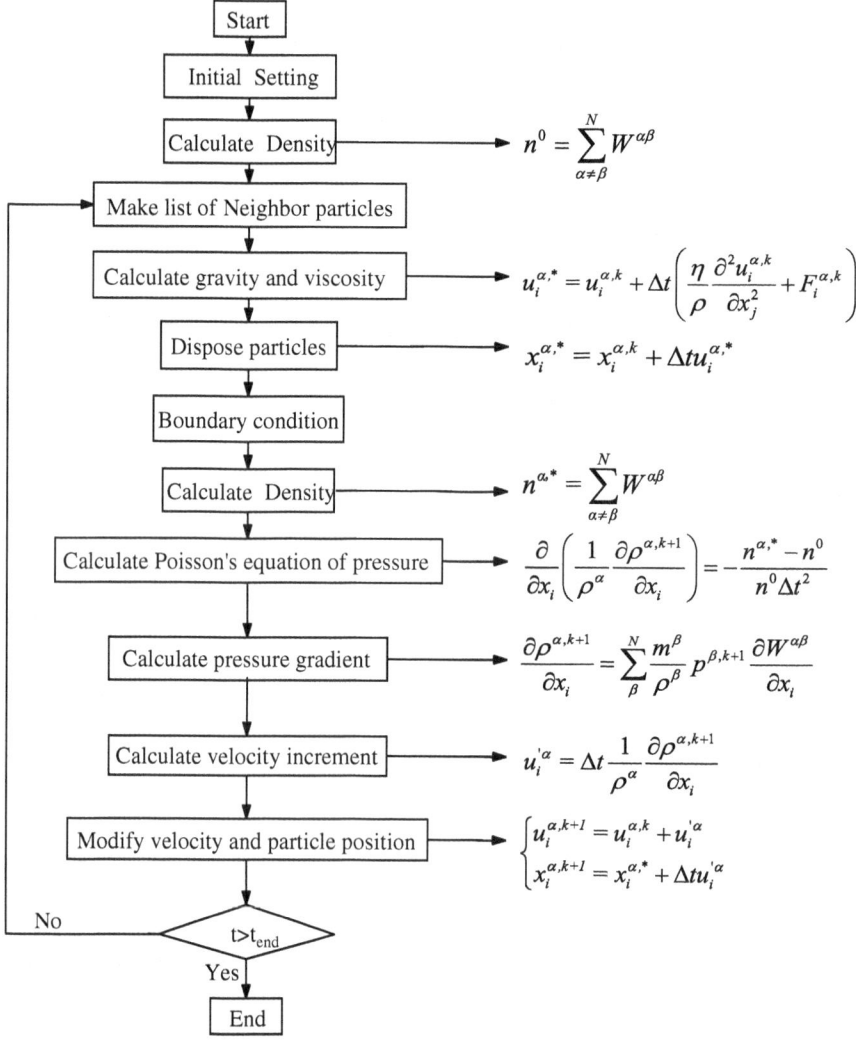

Fig. 3.21 Subroutine flowchart (reprinted from Huang et al. (2011) with permission from Springer)

the top and three textboxes and a group of radio buttons at the bottom (Fig. 3.22). The textboxes provide input windows for the number of particles to analyze. Radio buttons allow users to choose the type of information to display, such as velocity, displacement, or pressure.

First, the program defines dynamic arrays using the "ReDim Preserve" function, and then reads the data line-by-line from file STEP.DAT using a do-loop cycle and the "Line Input" function. The "Split" function splits the data and assigns them to corresponding dynamic arrays. Finally, the program expresses the

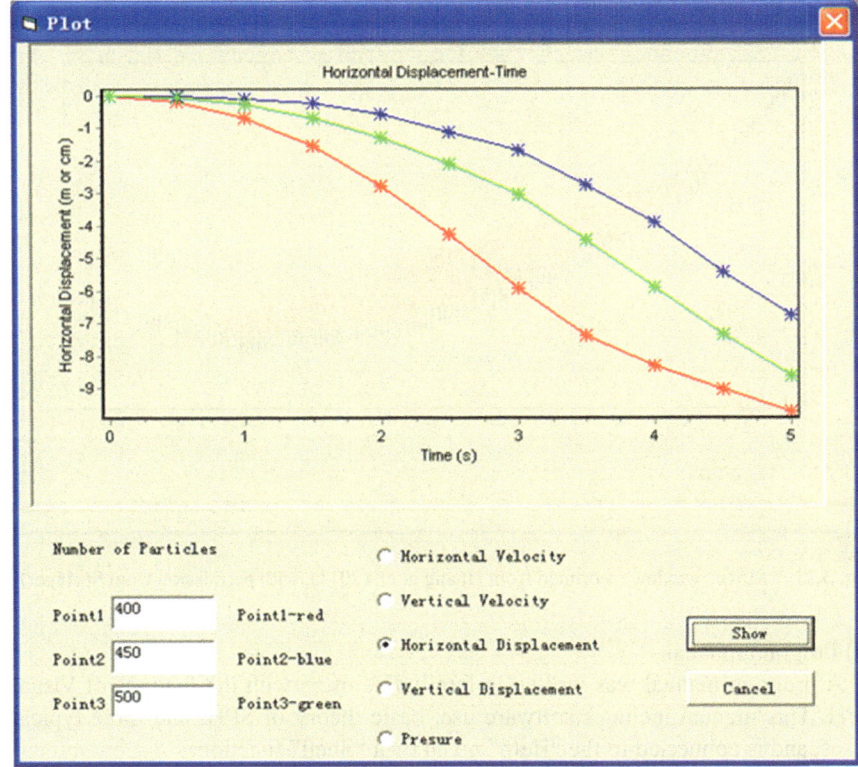

Fig. 3.22 Plot window (reprinted from Huang et al. (2011), with permission from Springer)

information stored in the arrays as curves in the picture frame, using the "Plot" function. Radio buttons are created by the "Option Button" function.

The aforementioned window changes complex computation results into a variety of graphic curves that precisely describe changes of velocity, displacement, and impact force during landslide movement, and facilitate understanding and analysis of the simulation results. Thus, the software provides an important basis for prediction and assessment of landslide hazards and the design of supporting structures.

Animation is executed mainly in the "Move" window. The principle of displaying dynamic graphics is similar to that for displaying static ones. The dynamic effect is achieved merely by refreshing the image from time to time. The program reads data from the file MOVE.MGF, splits these data and assigns them to corresponding dynamic arrays, and then draws the location of each particle. The latter step covers the previous steps, forming a continuous dynamic effect and displaying the process of landslide fluidization intuitively (Fig. 3.23). A control label is used to show the current step number, and the "Pause/Continue" button can be used to suspend animation at any time. Thus, it is convenient for users to analyze the movement and compare slope shapes with those observed in the field.

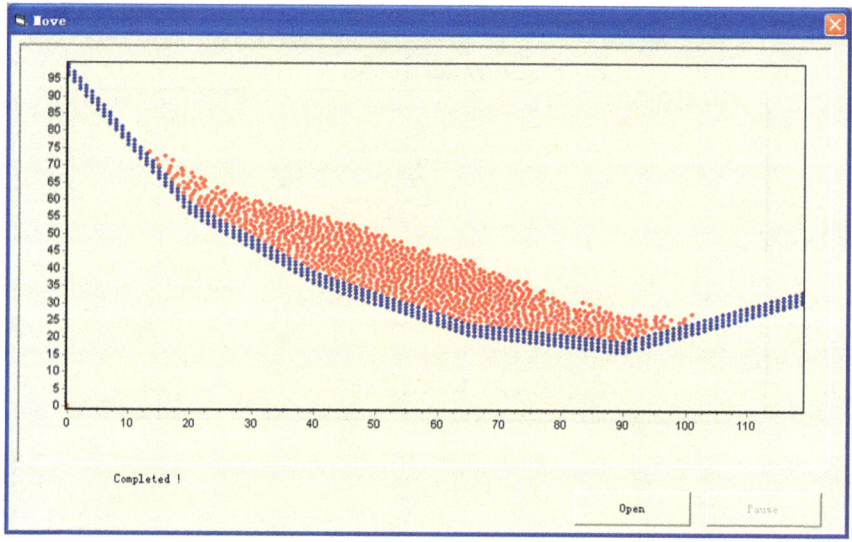

Fig. 3.23 "Move" window (reprinted from Huang et al. (2011), with permission from Springer)

(5) Program manual

A program manual was written to familiarize users with the features of Visual SPH. This manual includes software use, basic theory of SPH, and three typical cases, and is connected to the "Help" menu by a "Shell" function.

3.2.3 Application of the Software

Tangjiashan Mountain, on the right bank of the Tongkou River and 6 km upstream from Beichuan County, failed during the Wenchuan earthquake. The subsequent landslide was composed of weathered schist, slate, and sandstone sliding along rock. The difference in height between the landslide toe and the main back scar was 650 m, and the horizontal dimension of the landslide was 900 m (Hu et al. 2009). The landslide formed an extremely large barrier lake, with water storage capacity 250 million m³ (Cui et al. 2009). This lake is a serious threat to the city of Mianyang and other towns downstream.

By applying the software, a visual simulation was carried out to describe the fluidized movement of the Tangjiashan landslide. Parameters for the simulation were obtained from Hu et al. (2009) and are listed in Table 3.1.

Figure 3.24 shows that Visual SPH effectively reproduced the flow of earth materials. Users can observe landslide configurations at any time step. When comparing dynamic simulation results with surveyed landslide configurations, runout, slope coverage, and thickness from the software were very similar to the post-earthquake topographic map in Fig. 3.25. This map was taken from onsite

Table 3.1 Parameters used in SPH simulation of Tangjiashan landslide (reprinted from Huang et al. (2011), with permission from Springer)

Density	ρ (kg/m³)	2000
Equivalent viscosity coefficient	η (Pa·s)	1.9
Acceleration of gravity	g (m/s²)	9.8
Time step	n	1000
Unit time	$n\Delta t$ (s)	0.02

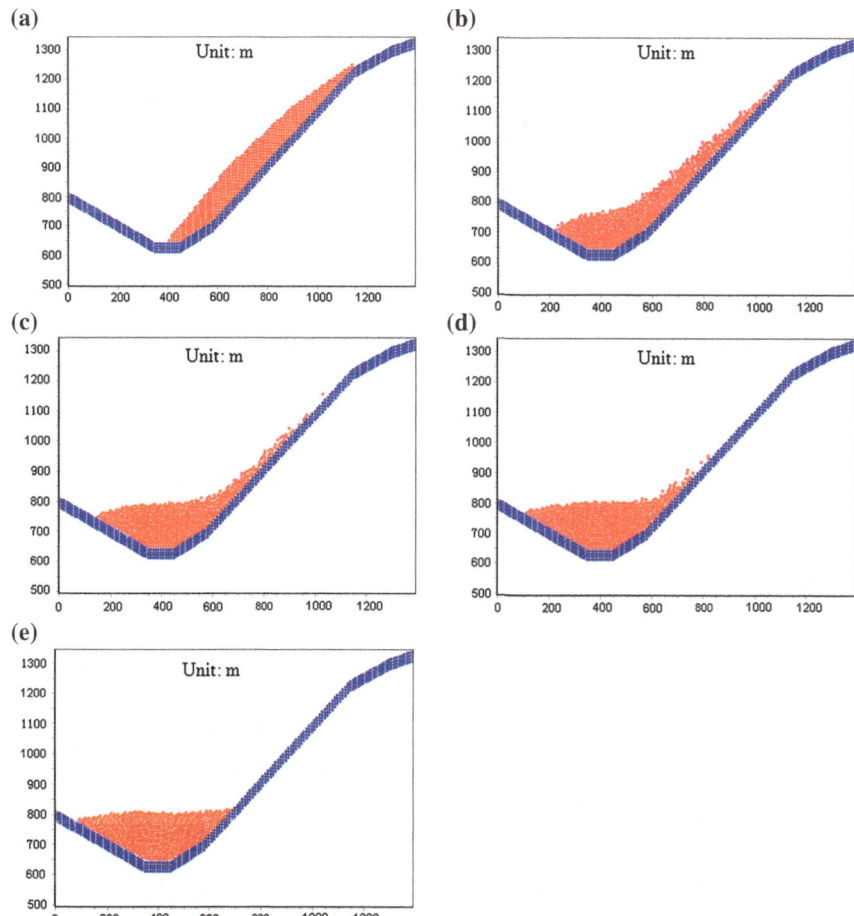

Fig. 3.24 Simulated runout process for Tangjiashan landslide (reprinted from Huang et al. (2011), with permission from Springer), **a** t = 0 s **b** t = 5 s **c** t = 10 s **d** t = 15 s **e** t = 20 s

measurements of Hu et al. (2009). The engineering geologic conditions in coverage have deteriorated significantly, and the mechanical properties of the soil materials are poor. Therefore, post-disaster reconstruction should be avoided in these areas. This confirms that the analysis results are useful in determining suitable locations for post-disaster reconstruction.

Fig. 3.25 Comparison of
SPH simulation and survey
data for Tangjiashan landslide
(reprinted from Huang et al.
(2011), with permission from
Springer)

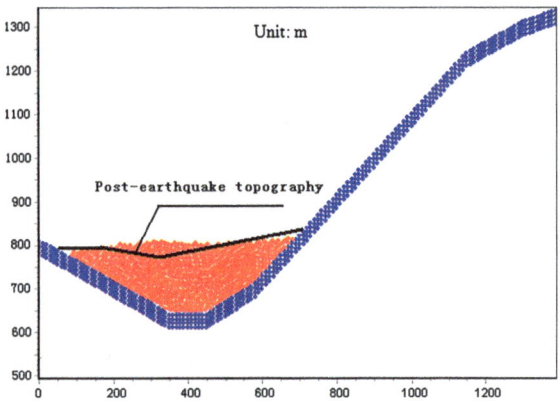

3.3 Summary

In this chapter, the means for implementation of the numerical model were
described. Detailed descriptions of each module were provided. In addition, visualization software based on the SPH model was developed with user-friendly interfaces. The main conclusions are as follows.

(1) According to the basic calculation process of the SPH method, an SPH calculation program that applies to large deformation simulation of soil material
was developed in a FORTRAN language environment. The function and flowchart of each module were detailed.
(2) Visualization software known as Visual SPH was successfully developed, using
a mixed programming approach. The software has high maneuverability and is
easy to use. User-friendly interfaces facilitate human–machine communication
and improve simulation efficiency. Static and dynamic graphic plots visualize the
abstract simulation results, making them easier to understand and analyze. The
software enables visual simulation for the analysis of landslide fluidized movement. The software can also accurately calculate velocity, impact force, and
other fundamental dynamic behaviors in the motion process, and can determine
essential landslide characterization parameters, including runout and coverage.

References

Hu, X. W., Huang, R. Q., Shi, Y. B., Lv, X. P., Zhu, H. Y., & Wang, X. R. (2009). Analysis of
blocking river mechanism of Tangjiashan landslide and dam-breaking mode of its barrier dam.
Chinese Journal of Rock Mechanics and Engineering, 28(1), 181–189. (in Chinese).
Huang, Y., Dai, Z. L., Zhang, W. J., & Chen, Z. Y. (2011). Visual simulation of landslide fluidized
movement based on smoothed particle hydrodynamics. *Natural Hazards, 59*(3), 1225–1238.
Cui, P., Zhu, Y. Y., Han, Y. S., Chen, X. Q., & Zhuang, J. Q. (2009). The 12 May Wenchuan
earthquake-induced landslide lakes: distribution and preliminary risk evaluation. *Landslides,
6*(3), 209–223.

Chapter 4
Validation of the SPH Models

In the last chapter, the Smoothed Particle Hydrodynamics (SPH) procedure based on hydrodynamics and elasto-plastic mechanics was developed for geo-disaster modeling. In this chapter, a series of validations for the SPH computer procedure are conducted. For the hydrodynamics SPH programs, the dam-break model and soil flow model tests are simulated, and the results are compared with those from the literature to determine the accuracy of SPH. For the SPH procedure in the framework of solid mechanics, a simple shear test of perfect elasticity and elasto-plastic material is simulated, and simulated stress–strain relationships are compared with the analytical solution and finite element method (FEM) results, thereby verifying the SPH models.

4.1 Dam-Break Model Test

4.1.1 SPH Simulation of Dam-Break Model Test

In this section we simulate a dam break, one of the classical free-surface problems in fluid dynamics, to verify and validate the accuracy of the SPH model. The initial model state is shown in Fig. 4.1. The water mass is colored blue and the boundary depicted in green. The left side of the water abuts the boundary, whereas the right side is free. Under the influence of gravity, the water simply flows from left to right along the fixed boundary at the base of the model. Model dimensions are shown in the figure. The height of the water mass is 0.2 m, and the width is 0.1 m.

The dam-break model of Fig. 4.1 is divided into a series of SPH particles. The total particle number N is 1,778, and the center distance of adjacent particles is 0.005 m. Water particles can move freely in both directions, but boundary particles are fixed. Parameters used in the simulation are shown in Table 4.1.

To visualize and analyze SPH simulation results, configurations of water mass at times $t = 0.10, 0.15, 0.20$, and 0.30 s are shown in Fig. 4.2.

The simulated water configurations at the four times coincided with the morphology observed in a dam-break model test. The velocity distribution at 0.15 s

© Springer-Verlag Berlin Heidelberg 2014
Y. Huang et al., *Geo-disaster Modeling and Analysis: An SPH-based Approach*,
Springer Natural Hazards, DOI 10.1007/978-3-662-44211-1_4

Fig. 4.1 Dam-break model

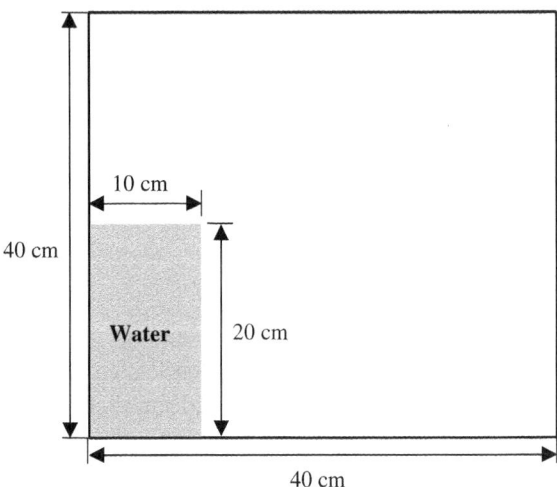

Table 4.1 Parameters used in SPH simulation of a dam break

Density	ρ (kg/m³)	1000
Equivalent viscosity coefficient	η (Pa·s)	1.7×10^{-3}
Acceleration of gravity	g (m/s²)	9.8
Time step	n	600
Unit time	Δt (s)	0.0005

is shown in Fig. 4.3. Velocities of the particles at the flow front are greater than those near the left boundary, and the maximum velocity is about 198.94 cm/s. The pressure distribution is depicted in Fig. 4.4. The pressure of the water particles increased with depth.

4.1.2 Verification of Dam-Break Numerical Model

Martin and Moyce (1952) conducted a water column collapse test and developed a scaling law for dam-break model testing. The scaling law included the following equations:

$$\frac{z}{a} = F_1[n^2, t(g/a)^t], \tag{4.1a}$$

$$\frac{\eta}{a} = F_2[n^2, t(g/a)^t]. \tag{4.1b}$$

where an^2 is the initial height of the water column;

a is the width of the water column;
z is the distance from the flow front to the coordinate origin;

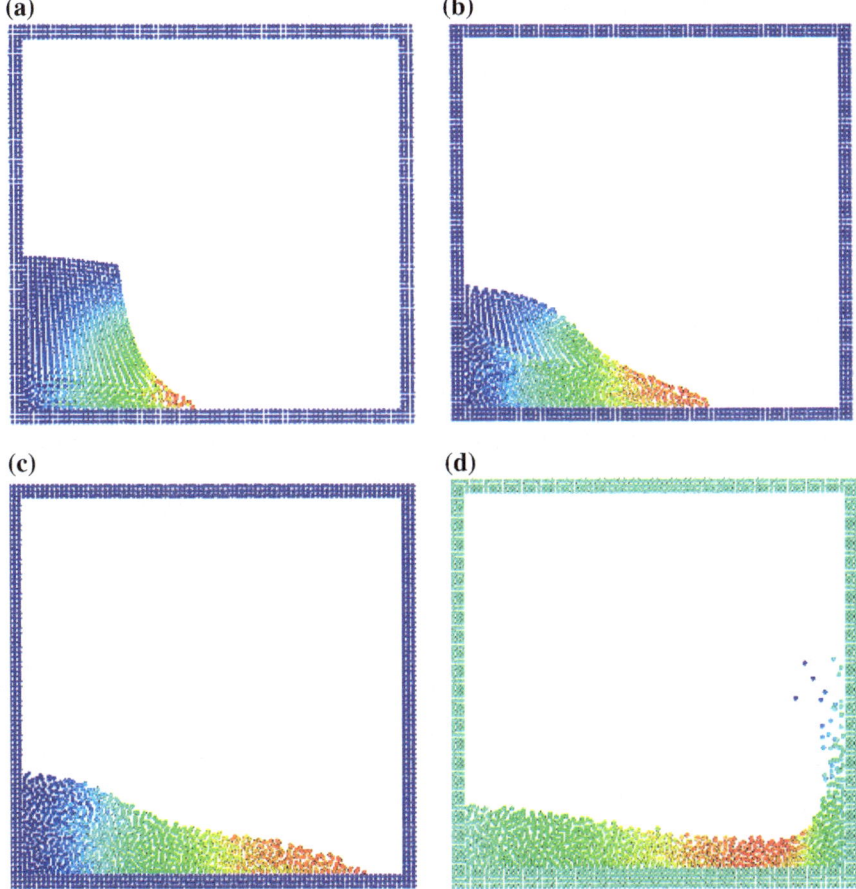

Fig. 4.2 SPH simulation results. **a** t = 0.10 s, **b** t = 0.15 s, **c** t = 0.20 s, **d** t = 0.30 s

η is the residual height of the water column;

g is the acceleration of gravity;

t is time.

For comparison's sake, Eqs. (4.1a) and (4.1b) were combined, and a series of standardized comparison equations were then proposed. These equations are now widely applied to compare simulation results and classical model test results for dam-break problems.

$$\begin{cases} Z = z/a; \\ H = \eta/an^2; \\ T = nt(g/a)^{\frac{1}{2}}. \\ \tau = t(g/a)^{\frac{1}{2}} \\ U = \mathrm{d}Z/\mathrm{d}T \end{cases} \qquad (4.1c)$$

Fig. 4.3 Velocity distribution of water mass at 0.15 s (unit: cm/s)

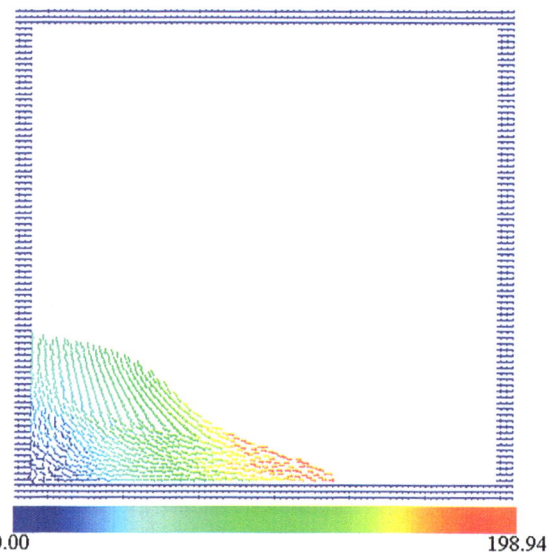

0.00   198.94

Fig. 4.4 Pressure distribution of water mass at 0.15 s (unit: kPa)

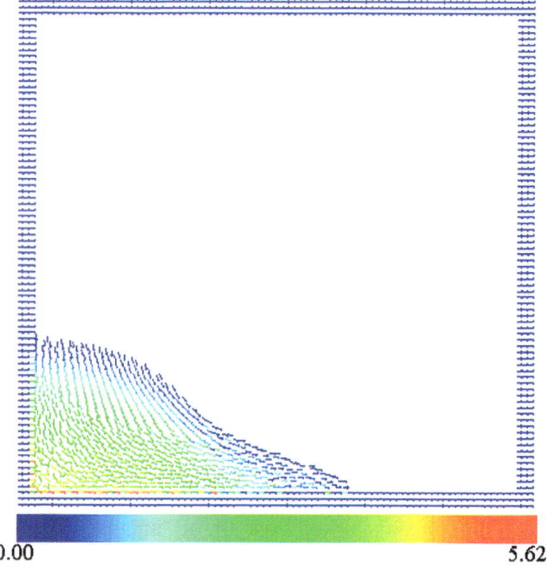

0.00   5.62

The scaling law is applied to the numerical modeling in this chapter. The height of the water $H = 0.2$ m, and its width $L = 0.1$ m, so n^2 is taken as 2. Then, the standardized time and surge front are $T = t \, (2g/L)^{-1/2}$ and $X = Z/L$, respectively. The SPH-simulated results are compared (Fig. 4.5) with those calculated by the above equations, using data from the literature (Martin and Moyce 1952). The SPH results agree with the test results, thereby demonstrating the accuracy of the SPH program.

Fig. 4.5 Comparison of SPH simulation and dam-break test (reprinted from Huang et al. (2011) with permission from Springer)

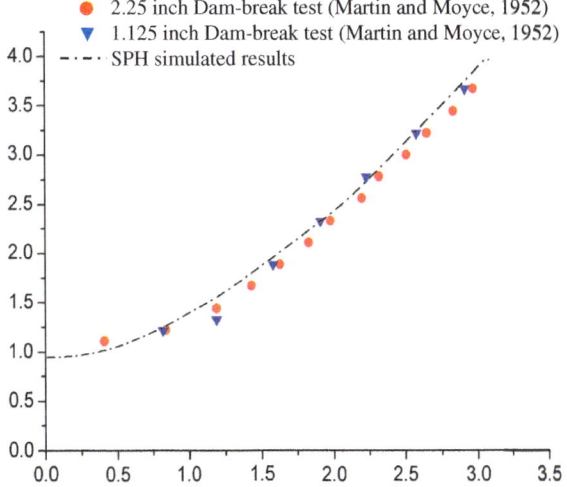

To further verify the accuracy of the SPH numerical modeling for the dam-break problem, the classical Ritter dam-break flood theory (Ritter 1892) was used to compare the SPH numerical results with the analytical solution. The channel is rectangular with a smooth, flat bottom. Downstream is dry with no water. The following equations can be derived from the Saint-Venant equations, using the theory of characteristic line (Zeng 2005):

$$h = \frac{1}{9g}(2\sqrt{gh_0} - \frac{x}{t})^2, \tag{5.2a}$$

$$u = \frac{2}{3}(\frac{x}{t} + \sqrt{gh_0})^2. \tag{5.2b}$$

where h is water depth; h_0 is the water depth of the upstream reservoir; t is time; x is the coordinate along the river channel, positive in the flow direction; and g is the acceleration of gravity.

It is easy to find that at position $x = 0$, water depth h, flow velocity u, and discharge per unit width q remain constant.

$$h = \frac{4}{9}h_0, \ u = \frac{2}{3}\sqrt{gh_0}, \ q = \frac{8}{27}h_0\sqrt{gh_0}. \tag{5.3}$$

The Ritter dam-break flood theory assumes that the water level upstream remains constant during the dam-break process. However, the above simulation is more similar to the collapse of a water column and does not meet the assumptions of the Ritter solution. Therefore, a new dam-break model (Quecedo et al. 2005; Fig. 4.6) is used to compare results with the Ritter solutions, in which the initial width of the water column is greatly increased to match the assumption.

Fig. 4.6 Dam-break model
in Ritter dam-break flood
theory

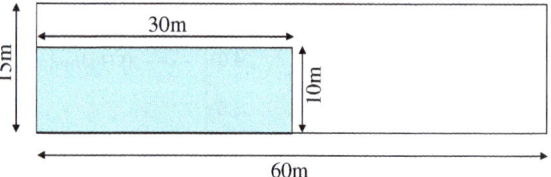

Fig. 4.7 SPH-simulated
results for water propagation
at various times. **a** t = 0.30 s,
b t = 0.70 s, **c** t = 1.00 s, **d**
t = 2.10 s

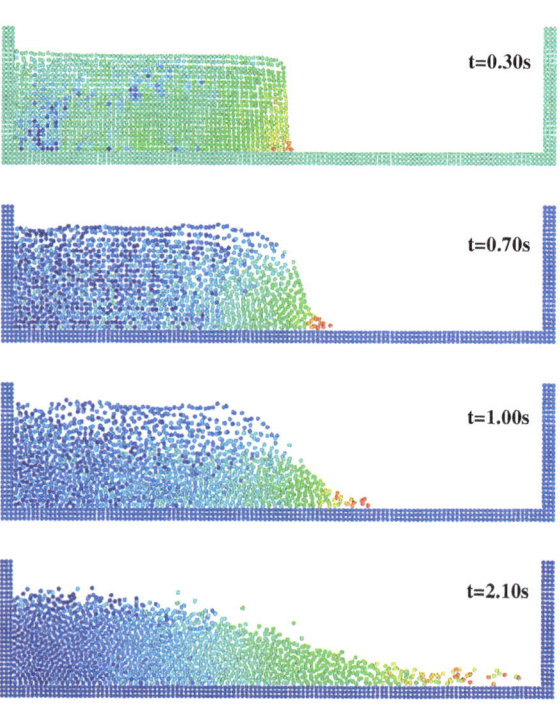

In the SPH discrete model, 1,830 SPH particles were used to represent water mass
and boundaries. The unit time step was 0.001 s and total time was 2.5 s (Fig. 4.7).

Figure 4.8 compares the simulated water configuration with the Ritter theory
analytical solution

Dynamic viscosity in the dam-break model was increased 1,000 times, and μ is
taken as 1.7 kg (ms)$^{-1}$. Then, the numerical simulation for a Newtonian fluid with
high viscosity can be calculated. Results at t = 0.10, 0.15, 0.20, and 0.30 s are por-
trayed in Fig. 4.9

Comparing the results with those in Fig. 4.2, the velocity of the fluid particles
was much less. Although the external configuration was essentially the same, the
distribution of fluid particles at every time point varied substantially. Under the
influence of high viscosity, the fluid particles became more compact during flow,
and the number of particles splashing at the flow front striker was significantly
reduced (Fig. 4.9).

(a)

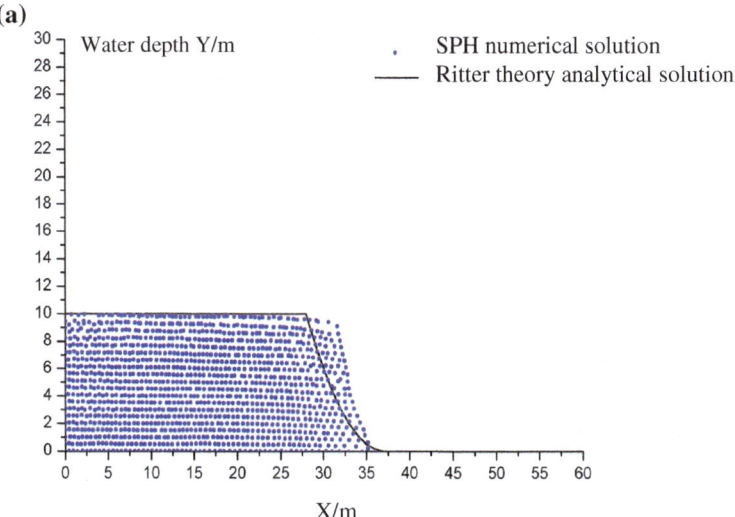

(b)

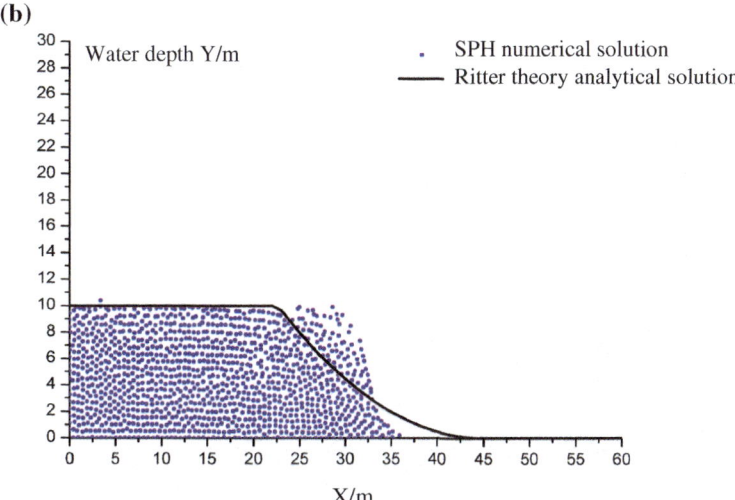

Fig. 4.8 Comparison of SPH simulation and Ritter analytical solution for dam break. **a** t = 0.30 s, **b** t = 0.70 s, **c** t = 1.00 s, **d** t = 2.10 s

(c)

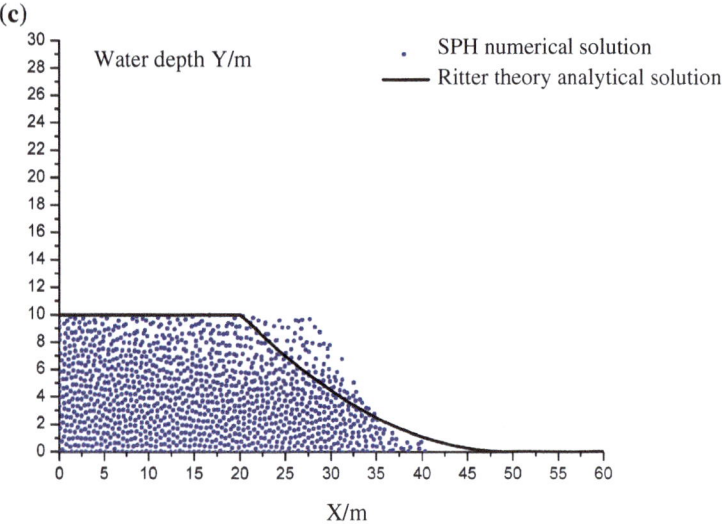

(d)

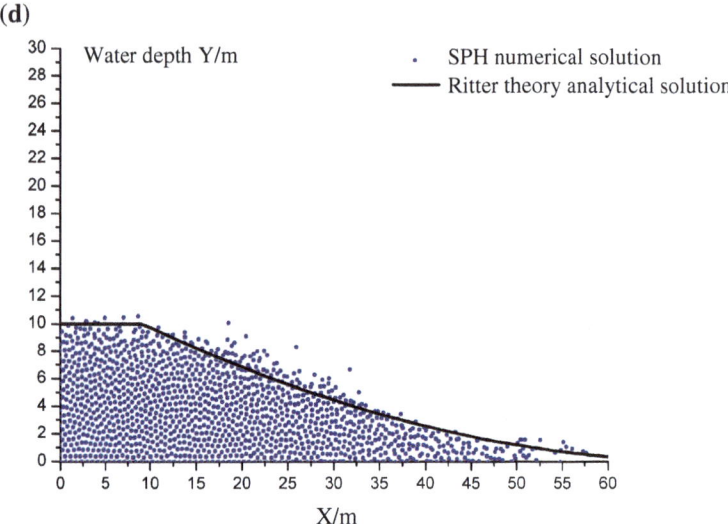

Fig. 4.8 (continued)

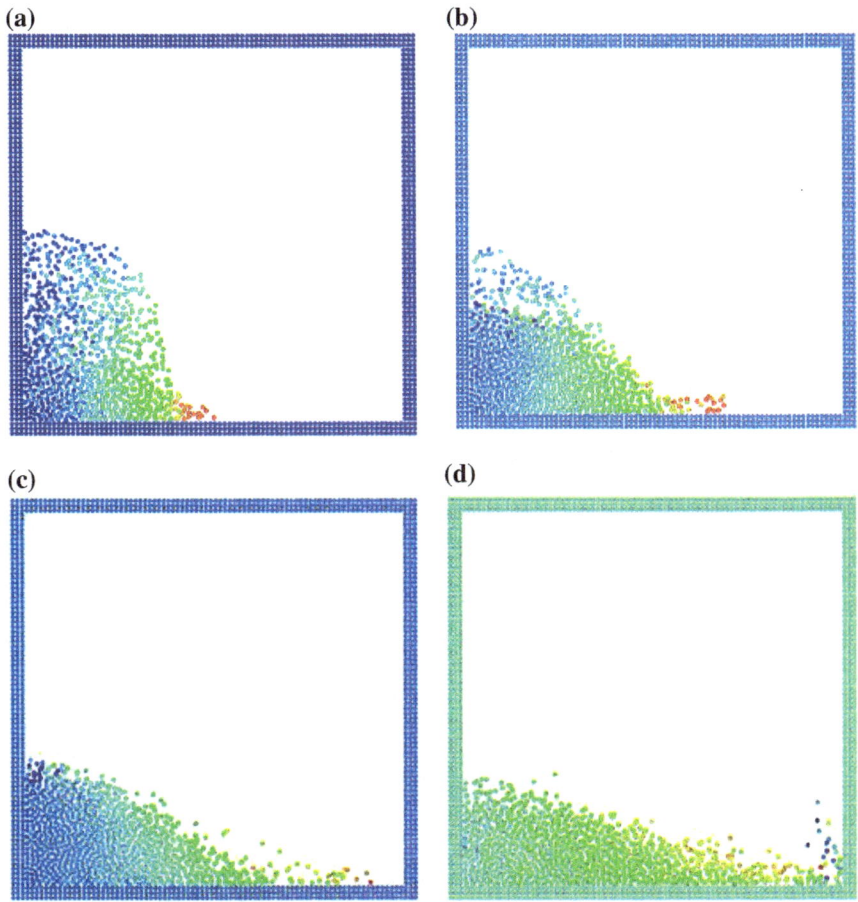

Fig. 4.9 SPH-simulated results for high-viscosity fluid dam break. **a** t = 0.10 s, **b** t = 0.15 s, **c** t = 0.20 s, **d** t = 0.30 s

4.2 Soil Flow Model Test

4.2.1 Bingham Fluid Constitutive Model

4.2.1.1 Flow Characteristics of Soil Material

The dynamic character of fluid in hydrodynamics can be described using

$$\tau = \eta_0\dot{\gamma}^n + \tau_y, \tag{4.4}$$

where τ is shear stress; $\dot{\gamma}$ is the rate of shear strain; and η_0, n, and τ_y are material parameters that vary with fluid type. Figure 4.10 shows various relationships between shear stress and shear strain.

Fig. 4.10 Various relationships between shear stress and shear strain (reprinted from Huang et al. (2012) with permission from Springer). **a** Newtonian fluid, **b** Bingham fluid, **c** pseudoplastic fluid, **d** dilatant fluid

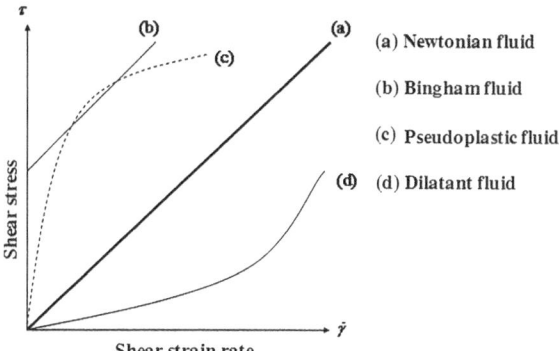

(a) Newtonian fluid

(b) Bingham fluid

(c) Pseudoplastic fluid

(d) Dilatant fluid

Fig. 4.11 Relationships between shear stress and shear strain for different fluid types. **a** Newtonian fluid, **b** Bingham fluid, **c** pseudo-plastic fluid, **d** dilatant flow

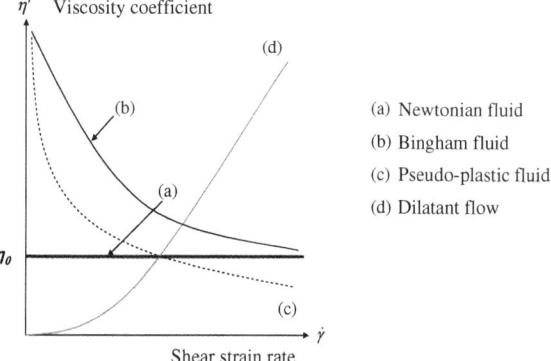

(a) Newtonian fluid

(b) Bingham fluid

(c) Pseudo-plastic fluid

(d) Dilatant flow

If $n = 1$ and $\tau_y = 0$, then the equation above represents the motion character of Newtonian fluid, and the constitutive relationship is represented by line (a) in the figure. Then, η_0 signifies the viscosity of the Newtonian fluid. If $n = 1$ and $\tau_y > 0$, the above equation represents the motion character of a Bingham fluid, and line (b) in the figure describes the constitutive law. Then, η_0 symbolizes the viscosity after yield. If $\tau_y = 0$ and $n > 1$, the fluid is expanding. If $n < 1$, the fluid is pseudoplastic. Relationships between the viscosity coefficient and shear strain rate are shown in Fig. 4.11.

In hydrodynamics, the difference between Newtonian flow and Bingham flow is as follows. The viscosity coefficient for Newtonian flow is invariant with shear strain rate, whereas this coefficient for Bingham flow varies with that rate.

In this chapter, the Bingham fluid model is used to simulate the large deformation of soil material:

$$\tau = \eta_0 \dot{\gamma} + \tau_y. \tag{4.5}$$

The following further explains how to calculate shear stress in a Bingham fluid in the SPH model and the corresponding set in large-deformation simulation of soil.

Fig. 4.12 Schematic of equivalent viscosity coefficient

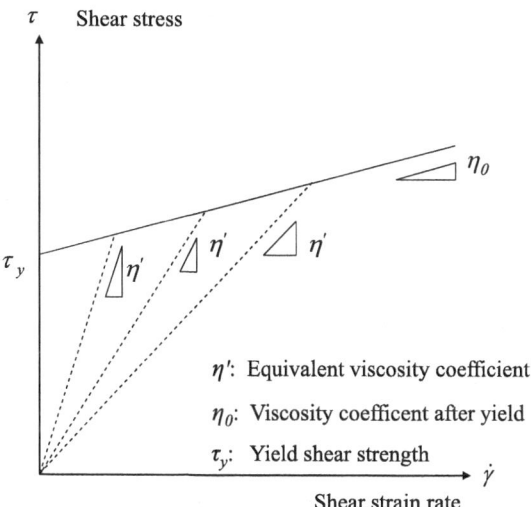

4.2.1.2 Equivalent Treatment for Viscosity Coefficient of Bingham Fluid

To better incorporate the Bingham fluid constitutive relationship into the SPH model, it is necessary to introduce the concept of equivalent Newtonian viscosity.

From Eq. (4.5), we can obtain

$$\eta' = \eta_0 + \frac{\tau_y}{\dot{\gamma}}, \tag{4.6}$$

where η' is the equivalent viscosity, τ is shear stress, $\dot{\gamma}$ is shear strain rate, η_0 is viscosity, and τ_y is the yield shear stress.

Figure 4.12 illustrates the concept of equivalent Newtonian viscosity. In this figure, the solid line represents the relationship between shear stress and shear strain rate in the Bingham model, whereas the dashed line represents motion characteristics expressed by the Bingham model with equivalent Newtonian viscosity. When the driving shear stress is smaller than the yield shear strength, soil behaves as a rigid body and does not deform. When the driving shear stress exceeds the minimum shear strength, soil behaves as a fluid, with a linear relationship to shear strain rate and considerable deformation.

To avoid an extremely large η', its maximum value was defined by Uzuoka et al. (1998):

$$\eta' = \eta_0 + \frac{\tau_y}{\dot{\gamma}} \quad when \quad \eta' < \eta_{max}$$

$$\eta' = \eta_{max} \quad when \quad \eta' > \eta_{max}, \tag{4.7}$$

where η_{max} is the maximum value of η'. Actually, in this form, motion characteristics of the soil material could be described using the Newtonian fluid model. When

Fig. 4.13 Dual linear model

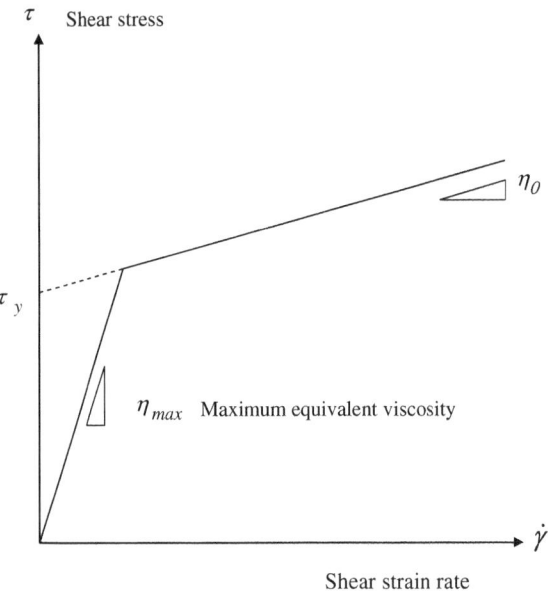

Fig. 4.14 Relationship between equivalent viscosity coefficient and shear strain rate

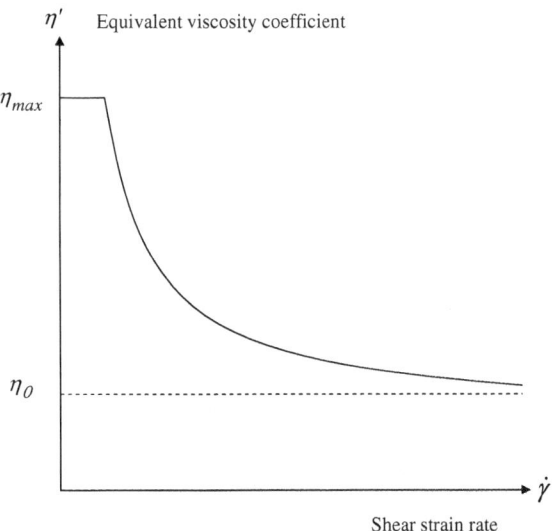

the shear rate is small, this material has a large η'. When the shear rate is large, the material has a relatively small η', and then this is a dual linear model (Fig. 4.13). The relationship between η' and shear strain rate is shown in Fig. 4.14.

η' promotes application of the Bingham fluid model to the large-deformation simulation of soil material. However, the Bingham model does not directly characterize material properties of the soil, so the Mohr-Coulomb failure criterion is introduced in the model. This criterion can be expressed by

$$\tau_y = \sigma_n \tan \varphi + c, \tag{4.8}$$

where τ_y is the shear stress of the soil material, σ_n is the normal stress, φ is the angle of internal friction, and c is cohesion. Substituting shear strength into Eq. (4.5), the Bingham model can be improved as follows.

$$\tau = \eta_0 \dot{\gamma} + \sigma_n \tan \varphi + c. \tag{4.9}$$

The final expression of η' in the Bingham model can be expressed as

$$\eta' = \eta_0 + \frac{\sigma_n \tan \varphi + c}{\dot{\gamma}}. \tag{4.10}$$

From the two-dimensional stress expression in the governing equations of hydrodynamics,

$$\sigma_{ij} = 2\eta_0 \dot{\gamma}_i - \frac{2}{3}\eta_0(\dot{\gamma}_i + \dot{\gamma}_j) = \frac{2}{3}\eta_0(2\dot{\gamma}_i - \dot{\gamma}_j). \tag{4.11}$$

Then, applying the concept of η', two-dimensional expressions of the viscosity in the Bingham model in different directions are

$$\begin{cases} \eta'_x = \eta_0 + \dfrac{4}{3}\eta_0 \tan \varphi + \dfrac{c - \frac{2}{3}\eta_0\gamma'_y}{\gamma'_x} \\ \eta'_y = \eta_0 + \dfrac{c}{\gamma'_y}. \end{cases} \tag{4.12}$$

4.2.2   SPH Simulation of the Soil Flow Model Test

The SPH method can accurately calculate kinetic parameters of particles, including velocity, displacement, and stress. Velocity and displacement were verified by the simulation for the dam-break problem. In this section, a model test of sand flow is simulated. Pressures of the soil particles and the impact force are calculated. The simulated results are compared with those of the soil flow model test.

The soil material was placed inside a model box, whose left side was near the left wall; the right side was free. Under the effect of gravity, the soil began to flow along the bottom of the box, and then impacted a baffle on the right side. The impact load was measured. The soil flow model tests were conducted with the model box placed at a different angle. A high-speed camera was used to record the soil configuration during flow. The camera recorded the soil configurations during flowing, and the impact force was measured at the right boundary (Fig. 4.15).

The temporal history of the impact force caused by soil flow with the model box at different angles is shown in Fig. 4.16.

Maximum impact forces on the baffle at different angles are listed in Table 4.2.

Fig. 4.15 Schematic illustration of slope model (reprinted from Huang et al. (2012) with permission from Springer)

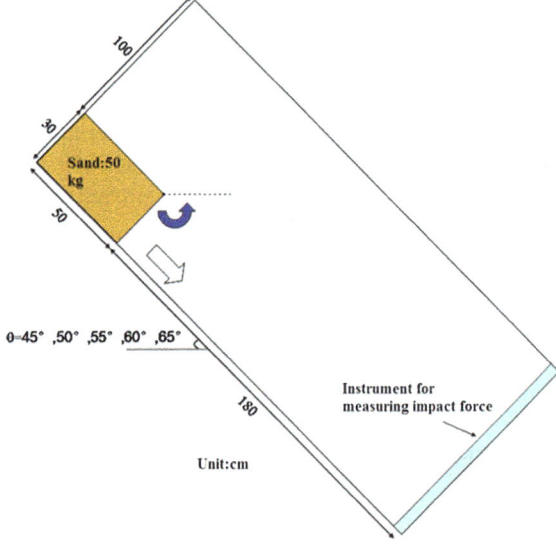

Fig. 4.16 Measured time series of impact force for different flume inclinations (reprinted from Moriguchi et al. (2009) with permission from Springer)

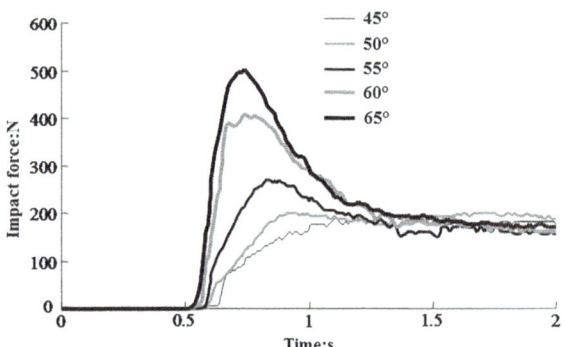

Table 4.2 Maximum impact forces on the baffle for different flume incinations (reprinted from Moriguchi et al. (2009) with permission from Springer)

Model box angle/	Maximum impact force/N				
	55	50	55	60	65
Test 1	168.9	212.7	269.0	395.1	500.5
Test 2	200.2	193.9	262.7	369.1	587.8
Test 3	187.7	200.5	256.5	506.6	512.9
Test 4	193.9	200.2	295.0	555.1	569.7
Test 5	212.7	206.7	300.2	337.7	512.9
Average	192.7	202.7	276.5	390.3	569.7
Standard deviation	15.5	6.5	17.5	35.7	16.6

The above model test was simulated by the SPH model. Soil parameters used in the simulation are listed in Table 4.3.

Table 4.3 Soil parameters used in the simulation

Density	ρ (kg/m^3)	1380.0
Viscosity coefficient	η_o (Pa·s)	1.0
Cohesion	c (kPa)	0.0
Internal friction angle	φ (°)	51.0

Fig. 4.17 Model box after discretization

Then, the model was discretized as in Fig. 4.17.

The model was divided into 11,339 particles, including 6,160 soil particles of radius 0.5 cm. The unit time step was 10^{-3} s and total duration 2 s. The smoothing length was set to 1 cm. Test and simulation results at different times are shown in Fig. 4.18.

Good agreement between the SPH simulation and model test of soil flow is seen, and therefore the accuracy of SPH can be verified and validated.

The SPH method can also calculate the pressure of each particle, and then the impact force of the soil flow applied on the right wall of the model box can be obtained. The time series records of impact force under slope angles of 45, 50, 55, 60, and 65° are shown in Fig. 4.19:

Compared with the test result (Fig. 4.16) under different slope angles, impact forces with time, their peak values and times of peak, and final stable values were all in good agreement.

The maximum impact forces, measured in the test and simulated, are plotted in Fig. 4.20. One can see that simulated and measured values are very close, and the increase with slope angle is consistent.

4.3 Slump Test for Viscous Material

In this section, SPH numerical simulation for a viscous material slump test is described. The viscous material was placed on a flat surface, and slumping occurred under the effect of gravity. The aim was to validate the SPH solution through comparison with results from the cubic interpolated pseudo-particle (CIP) method (Figs. 4.21 and 4.22).

Parameters used in the SPH simulation for the viscous material slump test are listed in Table 4.4. Five calculation conditions were used to fully verify the SPH model (Table 4.5).

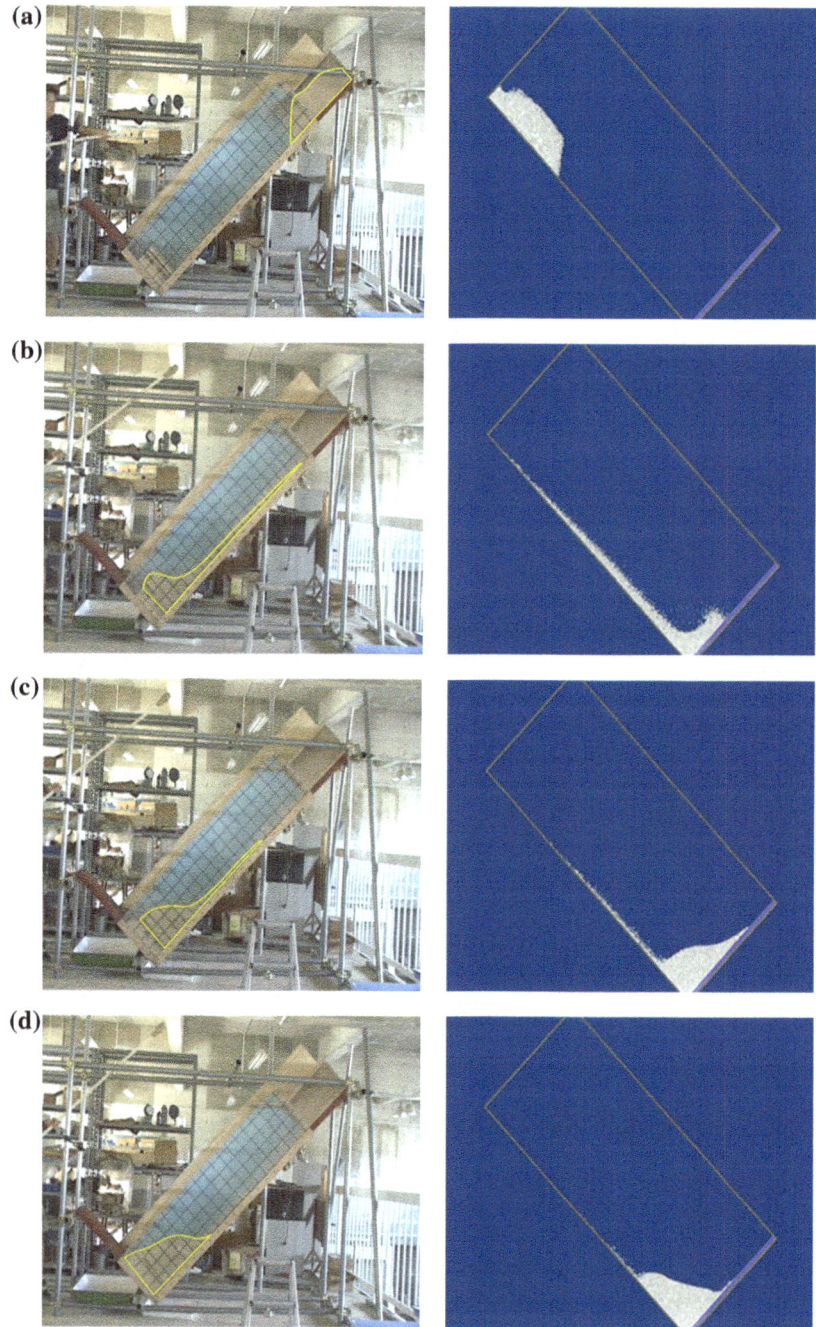

Fig. 4.18 Comparison of the results from the model test and SPH simulation (the test results are reprinted from Moriguchi et al. (2009) with permission from Springer). **a** t = 0.2 s, **b** t = 1.2 s, **c** t = 1.6 s, **d** t = 2.0 s

Fig. 4.19 Time series of impact force from SPH numerical simulation (reprinted from Huang et al. (2012) with permission from Springer)

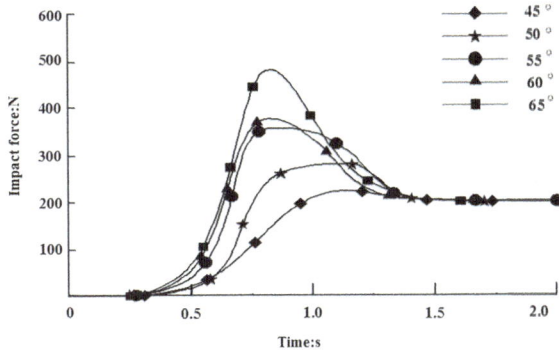

Fig. 4.20 Comparison of SPH simulation and test for granular flow (reprinted from Huang et al. (2012) with permission from Springer)

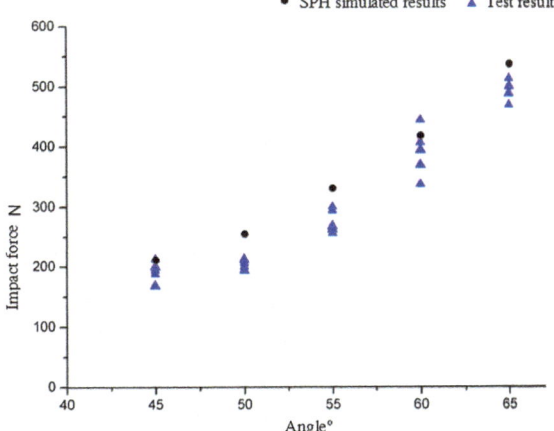

Fig. 4.21 Numerical model using SPH method

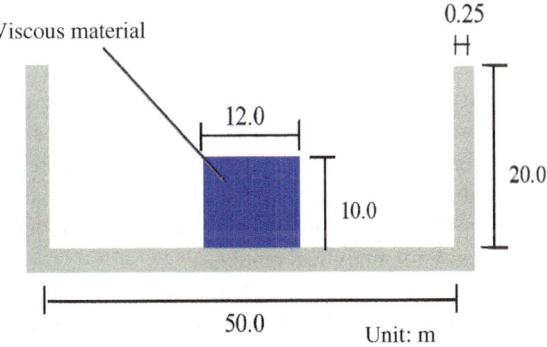

Fig. 4.22 Numerical model using CIP method (based on Moriguchi 2005)

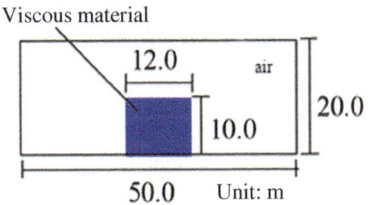

Table 4.4 Parameters used in SPH and CIP models (based on Moriguchi 2005)

	SPH method	CIP method
Material density ρ (kg/m^3)	2000.0	
Air density ρ (kg/m^3)		1.25
The speed of sound (through soil) v (m/s)		1500.0
The speed of sound (through air) v (m/s)		350.0
Time step Δt(s)	0.001	
Acceleration of gravity g (kg/m^3)	9.81	

Table 4.5 Calculation conditions

Case	1	2	3	4	5
Minimum viscosity coefficient η_0 (P$_a \cdot$ s)	1.0				
Cohesion c (Pa)	0.0	0.0	300.0	500.0	800.0
Angle of internal friction ϕ ($^\circ$)	30.0	55.0	0.0	0.0	0.0
Maximum viscosity coefficient η'_{max} (P$_a \cdot$ s)	1.0×10^8				

The viscous material was divided into 1,530 particles of radius 0.5 m. The smoothing length was 1 m. The time step was taken as 0.005 s, and the number of time steps was 500. The following gives simulation results from the SPH and CIP models. The configurations of the viscous material simulated by these two models (Figs. 4.23, 4.24, 4.25, 4.26 and 4.27) were consistent.

4.4 Simple Shear Test for Perfectly Elastic Solid

To verify the SPH model capacity for large-deformation analysis, a simple shear test was selected as a numerical experiment for easy simulation. To accurately compare simulation results with the theoretical solutions, the material was assumed a perfectly elastic solid, because there is only an analytic solution for the elastic model. The stress–strain relationship under shear stress was obtained and compared with analytical results.

A displacement control loading system was selected as the shear type for the simulation. To begin, an unstressed elastic solid was placed in a shear box. In a model of dimensions 300 × 300 mm (Fig. 4.28), a central region of 100 × 100 mm (blue square in Fig. 4.28) was considered the unit solid for calculation; the remaining part was the model boundary used to apply load. The displacement rate on the boundary is

$$V_x = 0.10\,y \ [cm/s], \tag{4.13}$$

where y is the distance from a particle to the horizontal axis.

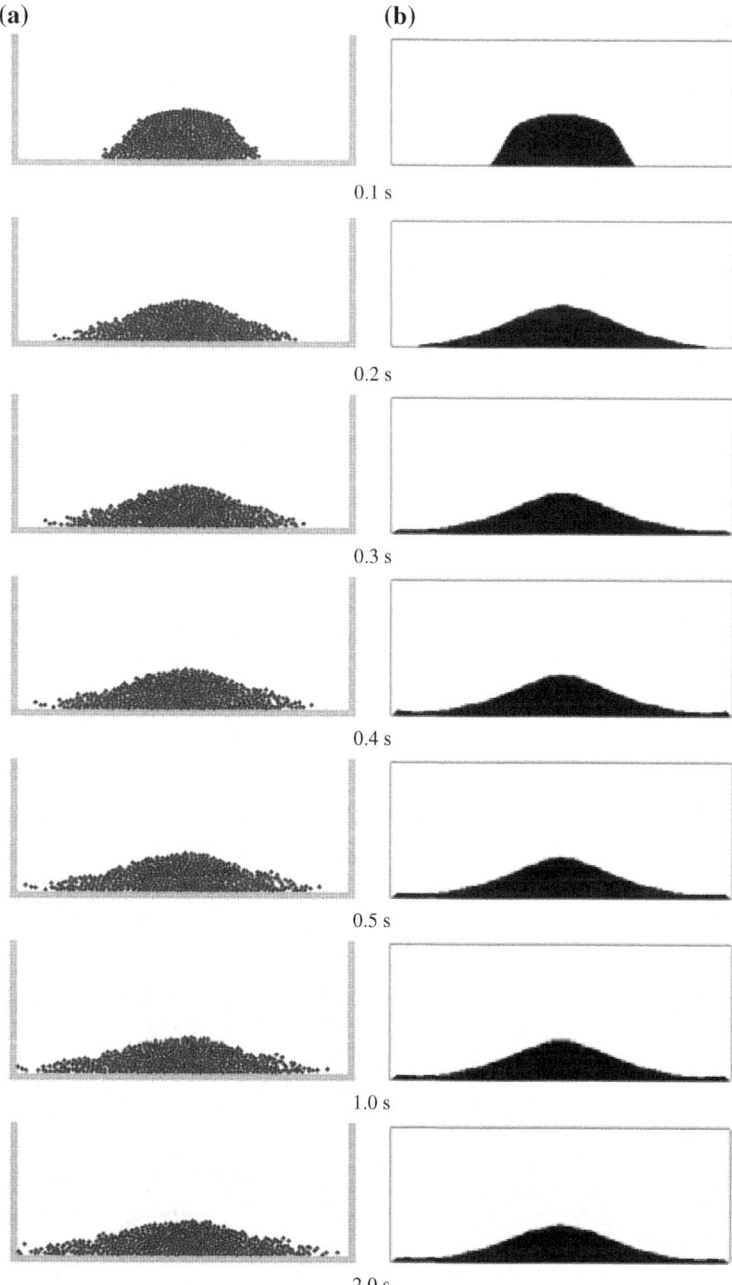

Fig. 4.23 Comparison of simulation results (*case 1*). **a** SPH simulation results, **b** CIP simulation results (based on Moriguchi 2005)

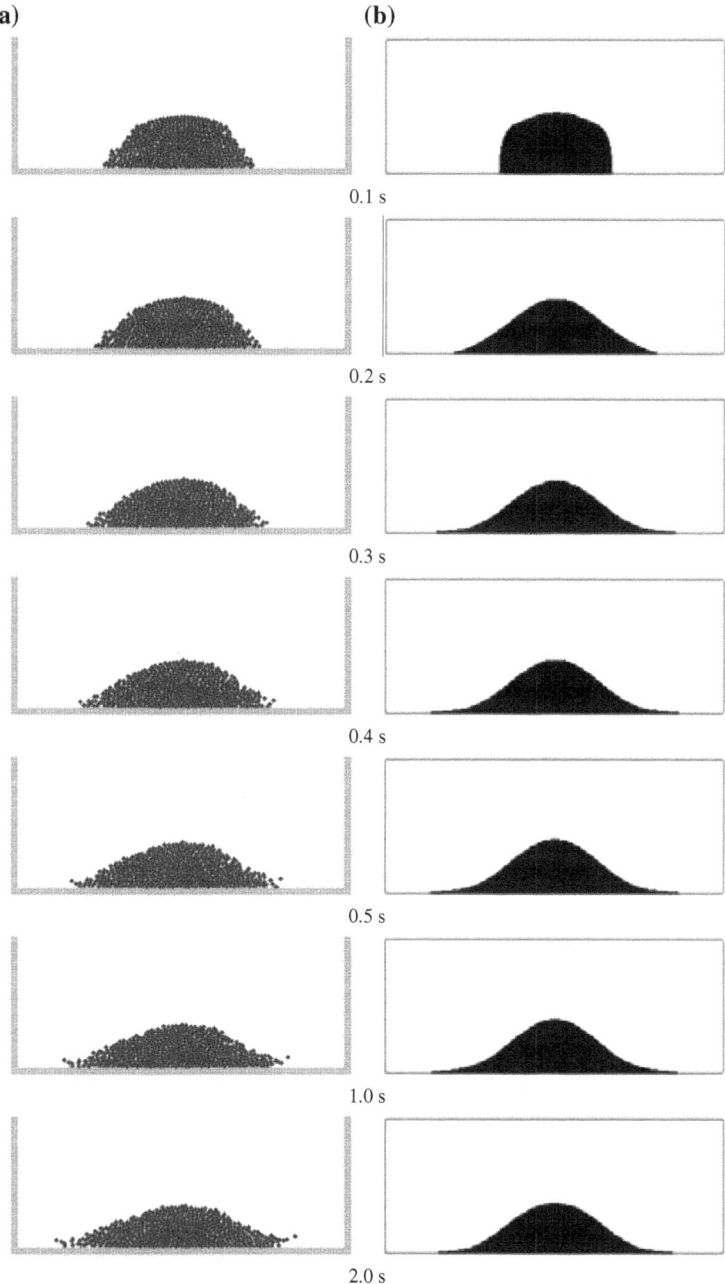

Fig. 4.24 Comparison of simulation results (*case 2*). **a** SPH simulation results, **b** CIP simulation results (based on Moriguchi 2005)

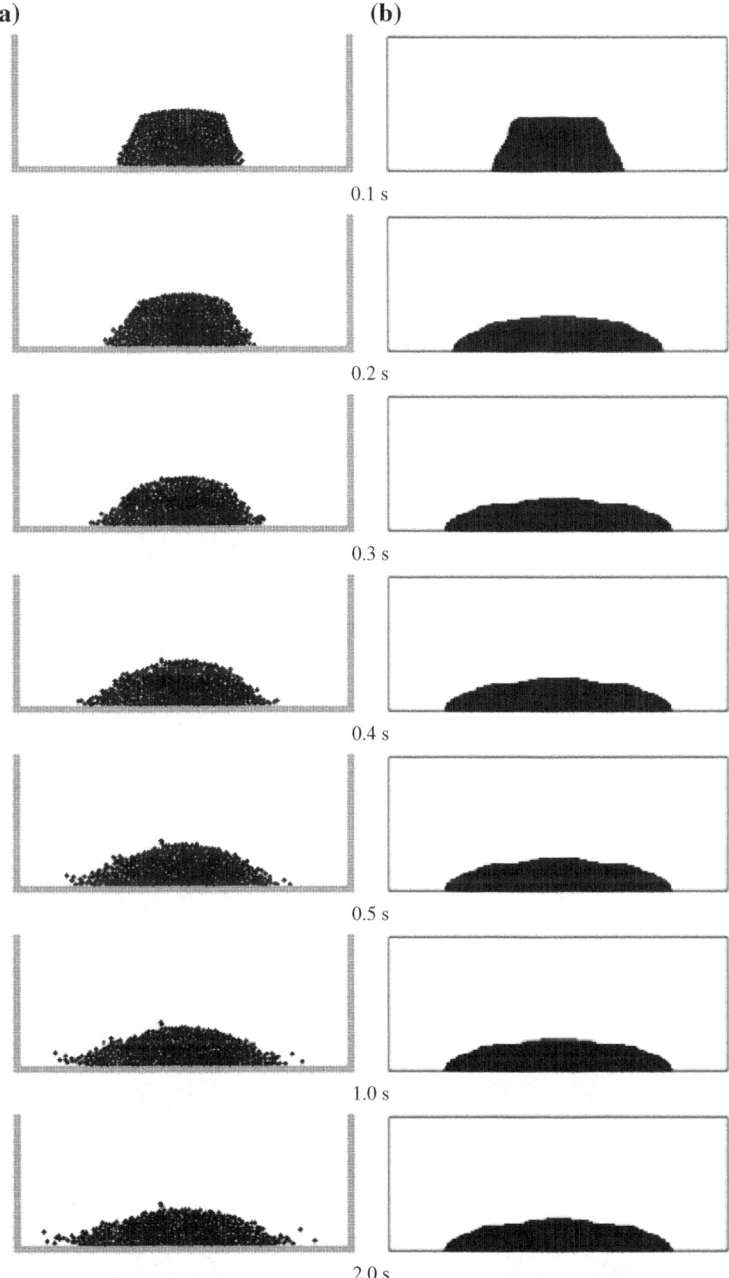

Fig. 4.25 Comparison of simulation results (*case 3*). **a** SPH simulation results, **b** CIP simulation results (based on Moriguchi 2005)

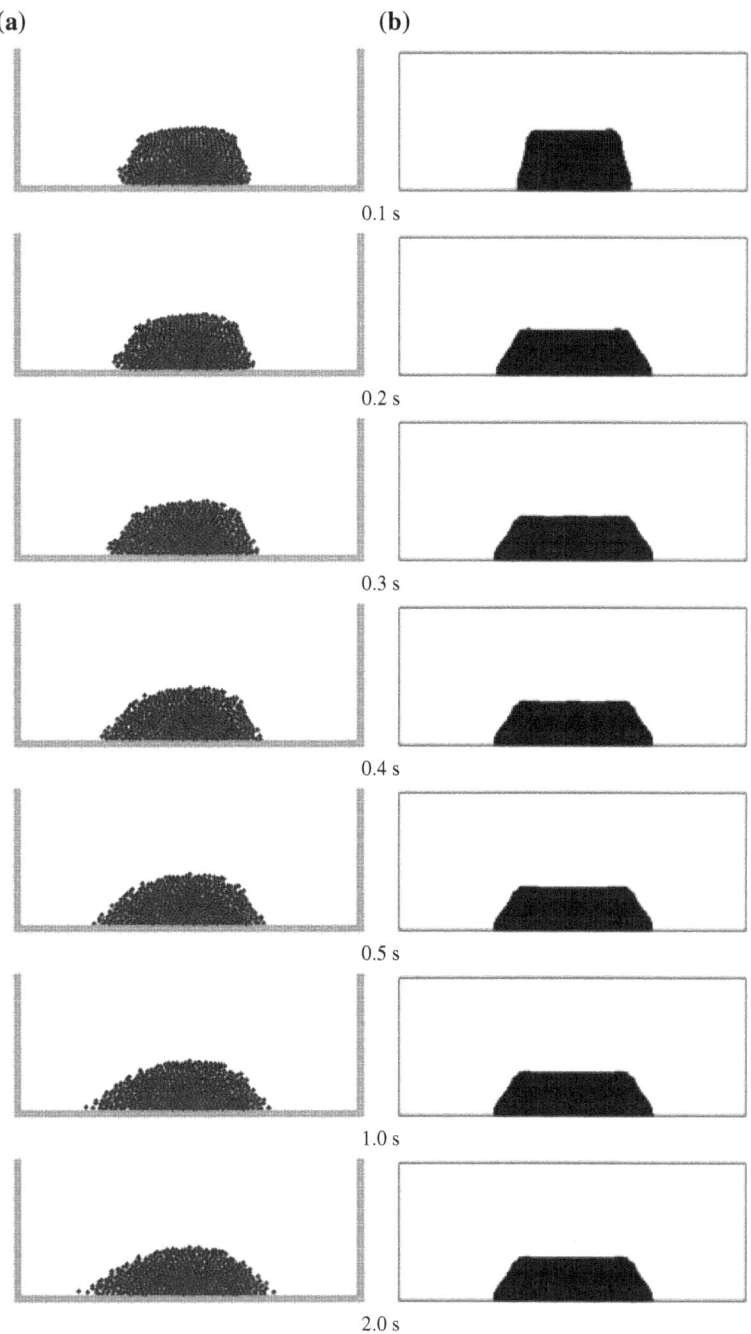

Fig. 4.26 Comparison of simulation results (*case 4*). **a** SPH simulation results, **b** CIP simulation results (based on Moriguchi 2005)

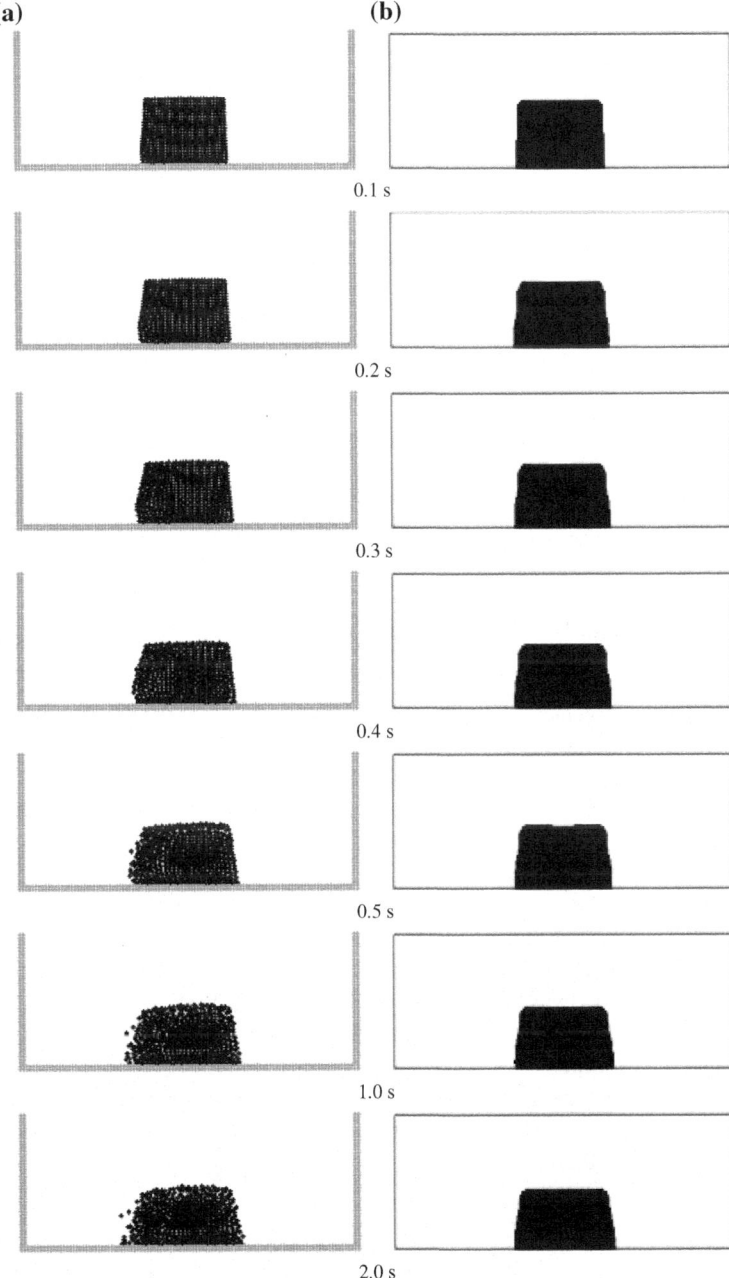

(a)   **(b)**

0.1 s

0.2 s

0.3 s

0.4 s

0.5 s

1.0 s

2.0 s

Fig. 4.27 Comparison of simulation results (*case 5*). **a** SPH simulation results, **b** CIP simulation results (based on Moriguchi 2005)

Fig. 4.28 Configuration
of elastic simple shear test
and its boundary conditions
(reprinted from Chen et al.
(2013) with permission from
Elsevier)

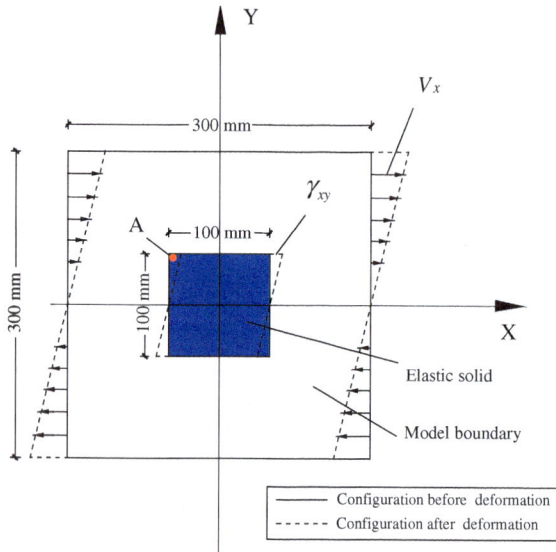

Table 4.6 Parameters in
SPH simulation of simple
shear (reprinted from Chen
et al. (2013) with permission
from Elsevier)

Young's modulus	E (kPa)	100.00
Poisson's ratio	ν	0.30
Density	ρ (kg/m³)	1600
Artificial viscosity parameter a, b	$a = b$	1.00
Artificial viscosity parameter k	k	0.01
Total steps	N	6000
Unit time step	ΔT (s)	0.001

The unit solid was assumed ideally elastic. Parameters in the SPH simulation
of the simple shear test are listed in Table 4.6. Analysis of the artificial viscosity
parameters a, b, and k was conducted by Liu and Liu (2003).

For the simple shear test described in Figs. 4.28 and 4.29 shows how the strain
state changed with time.

Figure 4.30 shows the stress state corresponding to the strain states in Fig. 4.29.

According to Hooke's law, the relationship between shear strain τ_{xy} and shear
stress γ_{xy} is described as

$$\tau_{xy} = G\gamma_{xy} = \frac{E}{2(1+\upsilon)}\gamma_{xy},\qquad(4.14)$$

where G is the shear modulus, E is Young's modulus, and υ is Poisson's ratio.

Fig. 4.29 Different strain states for simple shear test

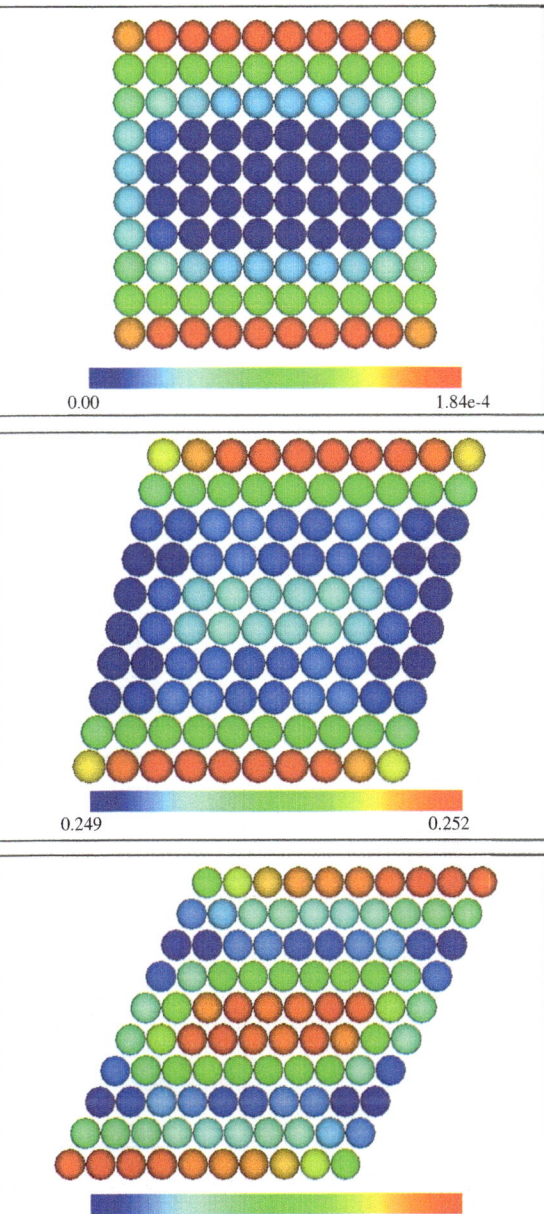

Fig. 4.29 (continued)

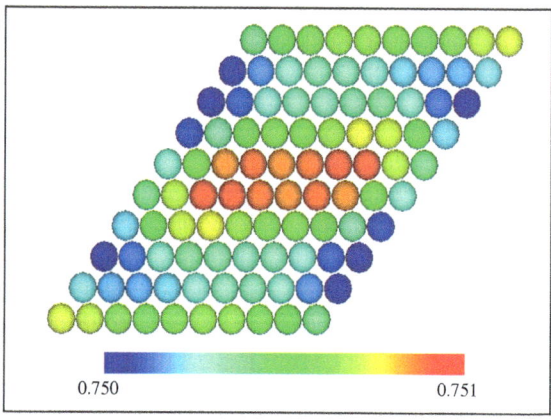

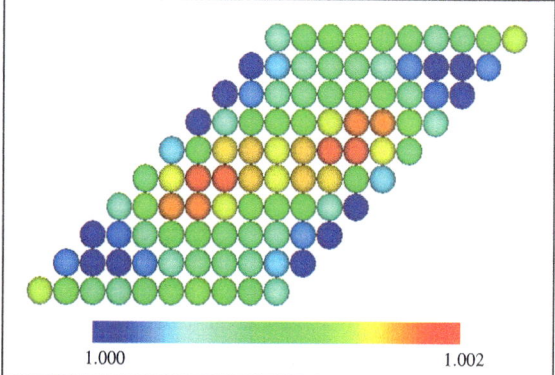

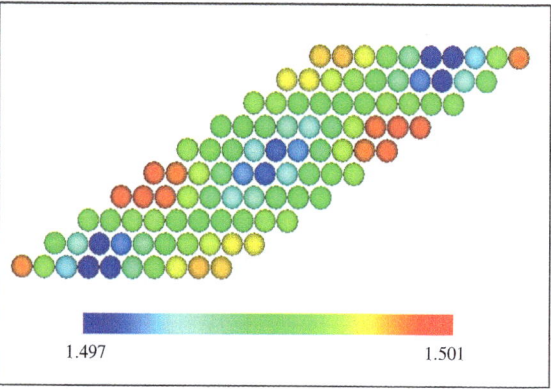

Fig. 4.29   (continued)

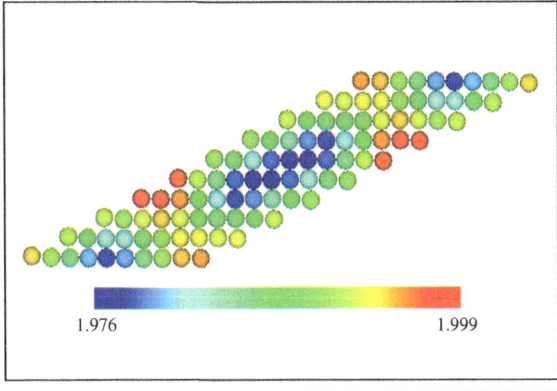

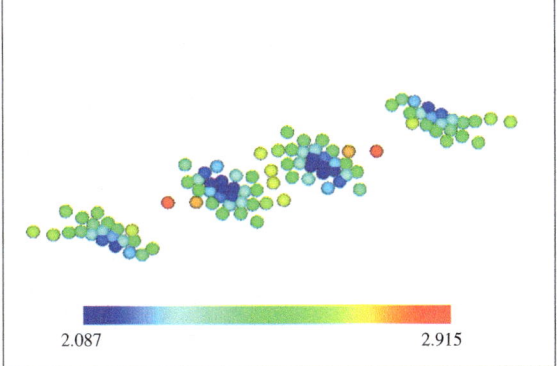

Fig. 4.30 Different stress
states for simple shear test
(*unit* kPa)

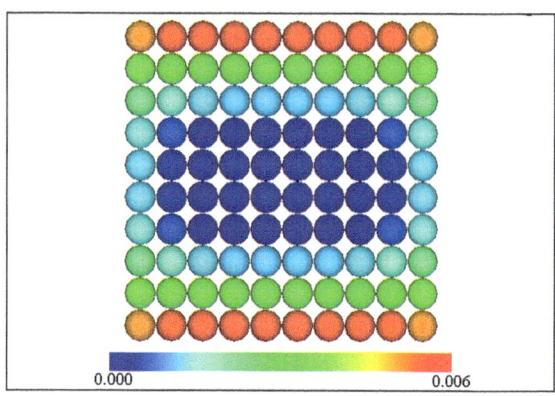

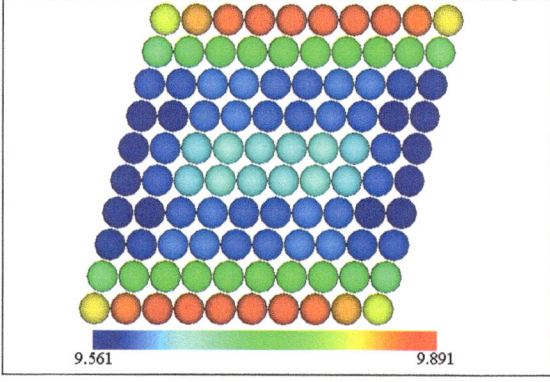

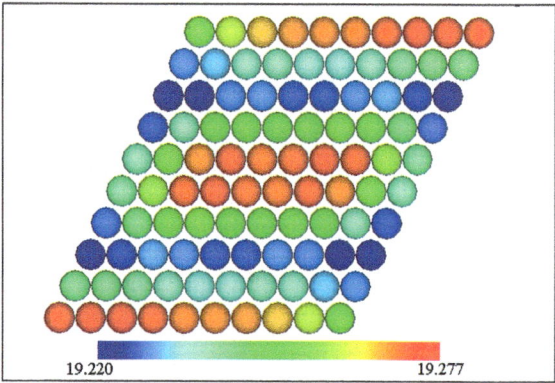

Fig. 4.30 (continued)

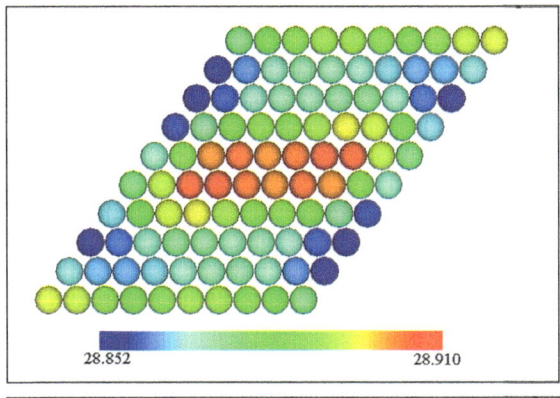

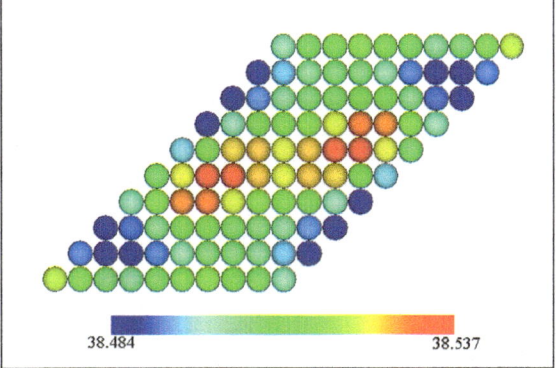

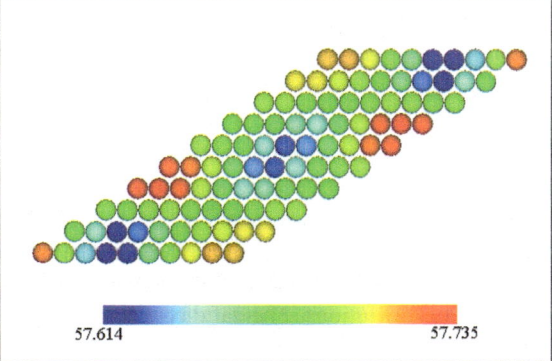

Fig. 4.30 (continued)

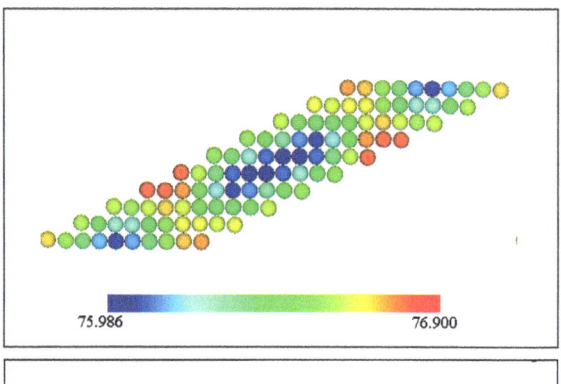

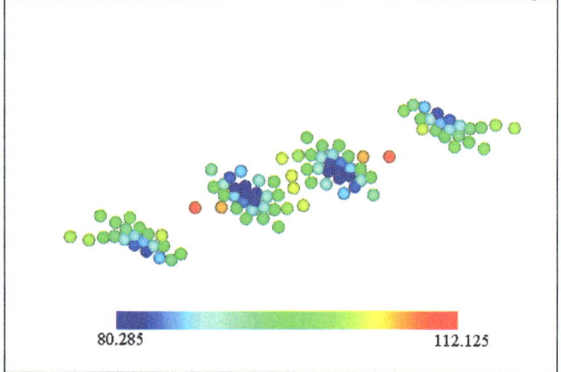

By substituting the parameter values in Table 4.5 into the equation, the analytical solution for the elastic simple shear test in this section can be expressed as

$$\tau_{xy} = 38.46\gamma_{xy}. \tag{4.15}$$

As an example, consider point A (Fig. 4.28) in the SPH numerical calculation as the analysis object. The comparison of its stress–strain relationship with the analytical solution is shown in Fig. 4.31. The black line is the analytical solution, whereas the red points show the calculations of the SPH model. The error is less than 1 %, which shows the goodness of fit between the simulation results and theoretical solutions. The validity and reliability of the SPH model are thereby confirmed.

4.5 Undrained Simple Shear Test for Soil

The simple shear test is a simple and effective way to study large soil deformation. In the experiment, soil samples are placed in a metal shear box. Normal pressure and shear stress are applied to make the soil fail along a specified shear surface.

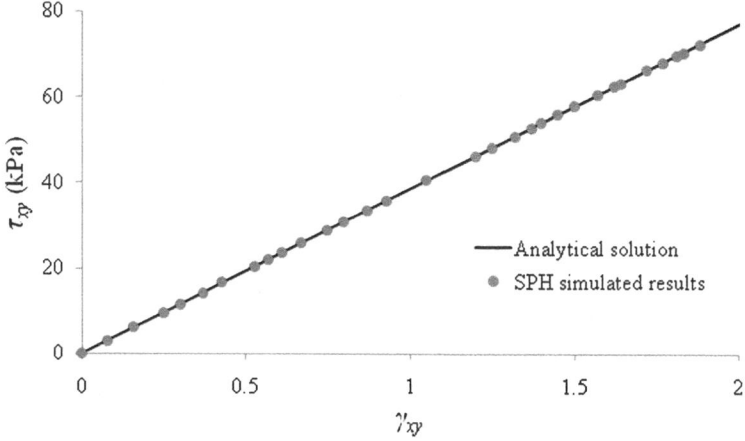

Fig. 4.31 Simulated and theoretical results of stress–strain relationship (reprinted from Chen et al. (2013) with permission from Elsevier)

Shear methods fall into two categories, stress-controlled and strain-controlled. For the stress-controlled shear instrument, horizontal stress is applied evenly to shear the soil sample. For the strain-controlled shear instrument, the shear box is pushed at a uniform speed and the soil sample is sheared with equal displacements. In this section, strain-controlled shear is simulated and a modified Cambridge model is used. Results from the SPH simulation are compared with the analytical solutions.

4.5.1 Cambridge and Modified Cambridge Models

The Cambridge and modified Cambridge models were proposed by Roscoe et al. (1963) at the University of Cambridge in the UK to describe the constitutive relationship of normal consolidation or poor-consolidation soil. Later, the models were popularized with consolidated clay, sand, and rock. Both models are isotropic hardening elasto-plastic models. The key assumption of these models is that soil deformation is only associated with three state variables, mean stress p, deviatoric stress q, and void ratio e.

Under the normal stress state, p is only associated with principal stresses σ_1, σ_2, and σ_3:

$$p = (\sigma_1 + \sigma_2 + \sigma_3)/3 \tag{4.16}$$

The q can be determined by

$$q = \frac{1}{\sqrt{2}}\sqrt{(\sigma_1 - \sigma_2)^2 + (\sigma_2 - \sigma_3)^2 + (\sigma_3 - \sigma_1)^2}. \tag{4.17}$$

Fig. 4.32 e-ln p curve

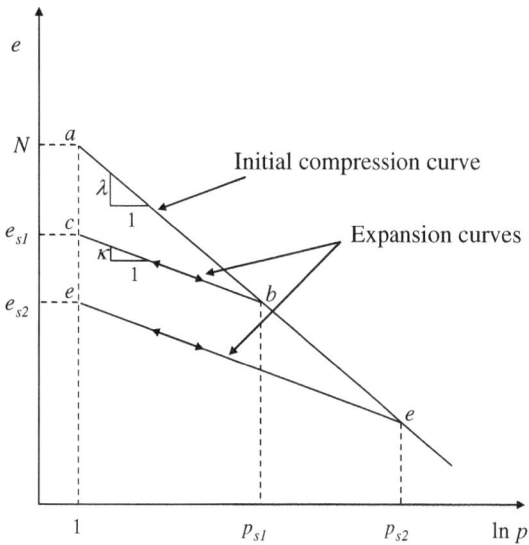

The Cambridge and modified Cambridge models were established based on important concepts, including the compression-rebound curve, critical state line, yield function, and others.

The models assume that when soft soil samples in a good drainage condition are slowly compressed under isotropic stress state conditions, an initial compression curve and a series of expansion curves are represented by the e-ln p curve (Fig. 4.32).

The initial compression curve can be determined by

$$e = N - \lambda \ln p. \tag{4.18}$$

Expansion curves can be expressed by

$$e = e_\mathrm{s} - \kappa \ln p. \tag{4.19}$$

Variables λ, κ, and N are feature parameters corresponding to a specific soil material. λ and κ are respectively slopes of the compression curve and expansion curve within the e-ln p curve. N is the void ratio corresponding to the unit pressure in the initial compression curve. As shown in Fig. 4.32, void ratios e_s vary with expansion curve and depend on the stress history of the soil. If the stress state of the soil sample is on the initial compression curve, then that soil is normal consolidation soil. If that stress state can be described using the curve bc when unloading, then the soil is overconsolidation soil.

A soil sample will eventually reach a state when subjected to a sustained shear stress in which the shear strain continues to increase while the stress or volume remains constant. That means that in this critical state, the soil sample enters a

Fig. 4.33 Critical state line (CSL) with normal consolidation curve

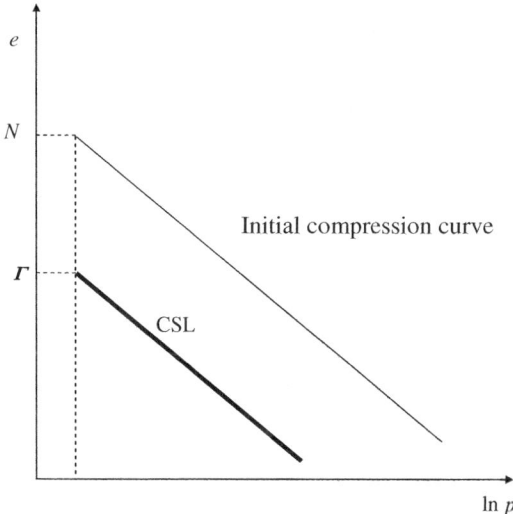

stage of steady deformation. This state is described by a critical state line (CSL). The CSL corresponding to normal consolidation soil is shown in Fig. 4.33. In e-ln p space, the CSL is parallel to the normal consolidation curve. From Eq. (4.19), the relationship between parameters N and Γ for the Cambridge model is

$$\Gamma = N - (\lambda - \kappa). \tag{4.20}$$

For the modified Cambridge model, the relationship between these two parameters is

$$\Gamma = N - (\lambda - \kappa) \ln 2. \tag{4.21}$$

After the constitutive model is determined, when applying q the soil is in the elastic stage before reaching the yield q. The value of the yield stress can be obtained using the yield equation, as follows.

For the Cambridge model:

$$q + Mp \ln(\frac{p}{p_0}) = 0. \tag{4.22}$$

For the modified Cambridge model:

$$\frac{q^2}{p^2} + M^2 p(1 - \frac{p_0}{p}) = 0. \tag{4.23}$$

In p–q space, the stress–strain relationship in the Cambridge model is a logarithmic curve. In the modified Cambridge model, it is an oval-shaped curve (Fig. 4.34). p_0 is the average principal stress value at the yield point. It determines

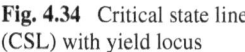

Fig. 4.34 Critical state line (CSL) with yield locus

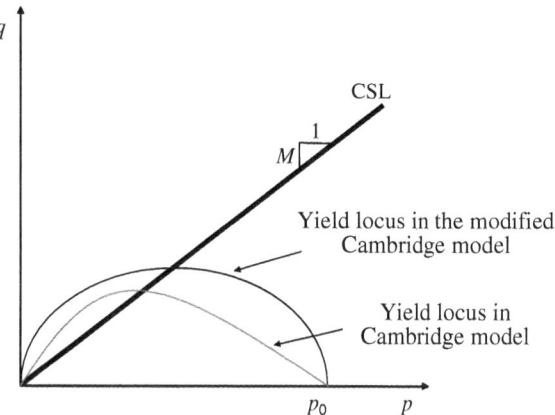

the form of the yield surface, and its value varies with the expansion curve. M is the slope of the critical failure line in p–q space. The intersection of the yield curve and critical line is the maximum value of q.

The parameters E, G, and υ are normally used in the Cambridge and modified Cambridge models. Generally, only two of these are needed in the analysis, and the other parameter can then be obtained. In this section, the modified Cambridge model is used to simulate the soil non-drained shear tests in the plane strain condition.

4.5.2 Numerical Simulation Results

The numerical model is similar to the elastic simple shear test model of the previous section. The strain increases with constant amplitude under undrained conditions. The initial state is such that the soil samples are placed in the shear box with no principal stress. The computation parameters used in this simulation example are listed in Table 4.7.

The simulated strain variation is shown in Fig. 4.35.

Corresponding to the strain variation in the above figures, simulated stress variation is shown in Fig. 4.36. The unit for these figures is kPa.

Table 4.7 Parameters used in SPH modeling with modified Cambridge model

Density ρ (kg/m^3)	Time step Dt (sec)	Artificial viscosity coefficient $a^{vis} = b^{vis}$	Artificial viscosity coefficient e^{vis}	Poisson's ratio υ	Compression index λ	Swelling index κ
1800	0.00010	5.00	0.01	0.33	0.355	0.0577
Initial void ratio e_0	Principal stress ratio R_f	Reference stress p_0 (kPa)	Initial average stress p (kPa)	Particle number	Particle spacing (cm)	Influence radius (cm)
1.55	5.12	100	100	900	1.00	2.00

Fig. 4.35 Simulated strain
variation when $\gamma = 1.18{-}1.21$

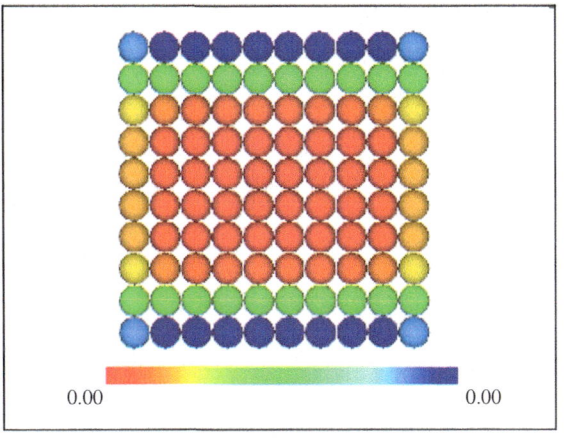

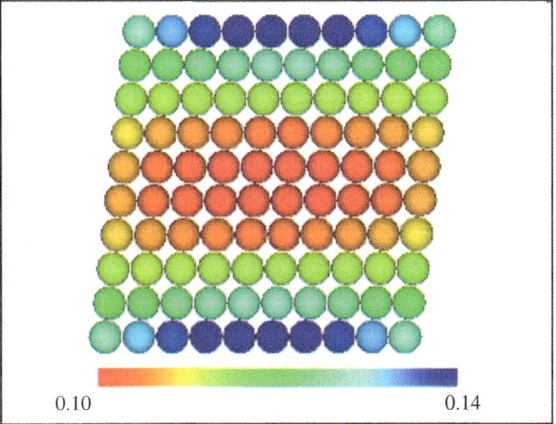

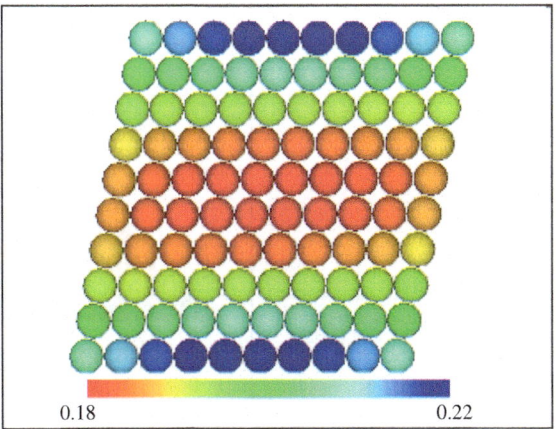

Fig. 4.35 (continued)

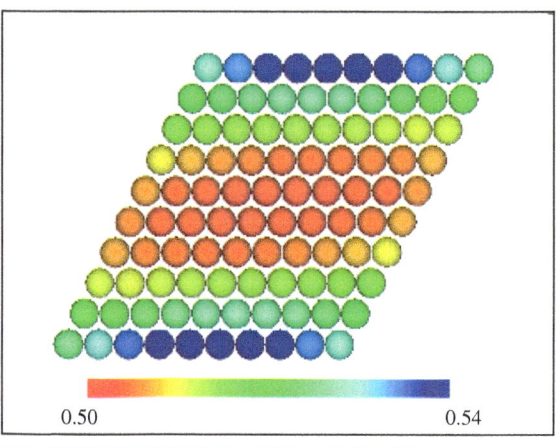

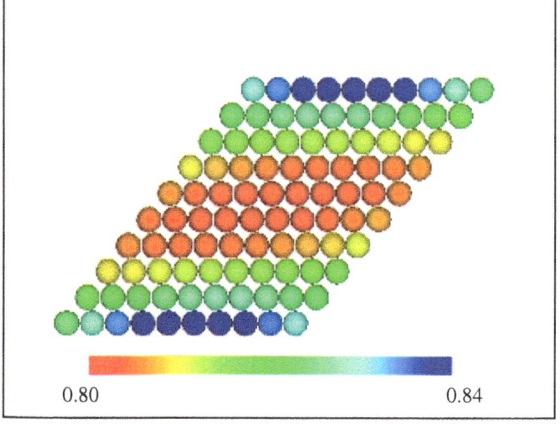

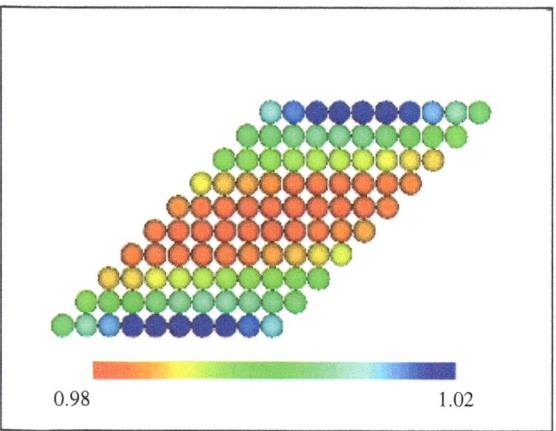

Fig. 4.35 (continued)

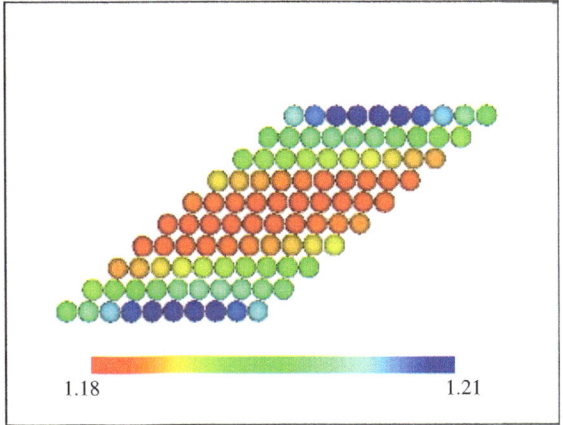

Fig. 4.36 Simulated stress variation when $\gamma = 1.18$–1.21

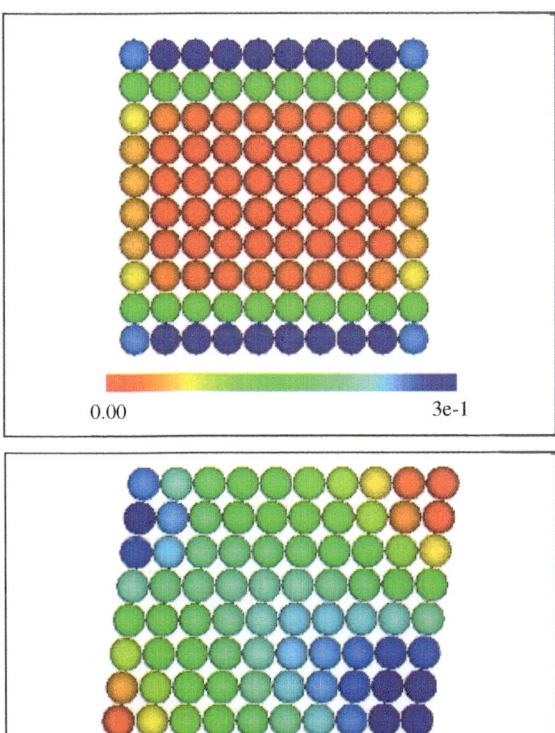

Fig. 4.36 (continued)

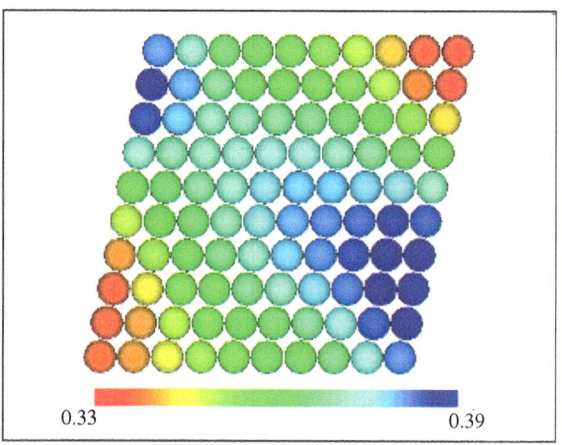

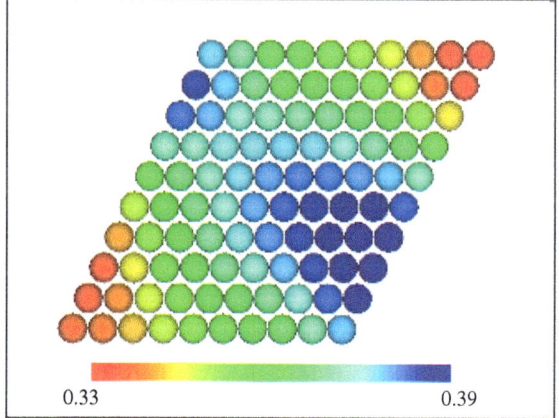

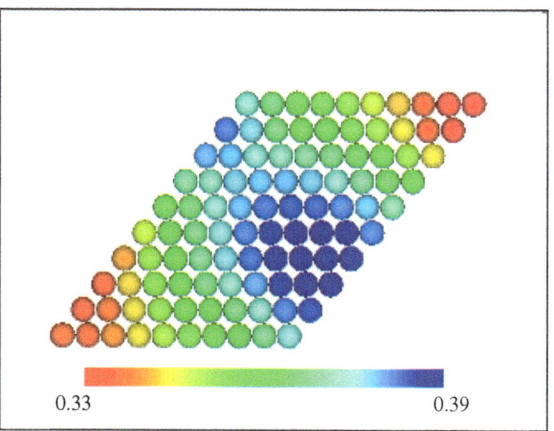

Fig. 4.36 (continued)

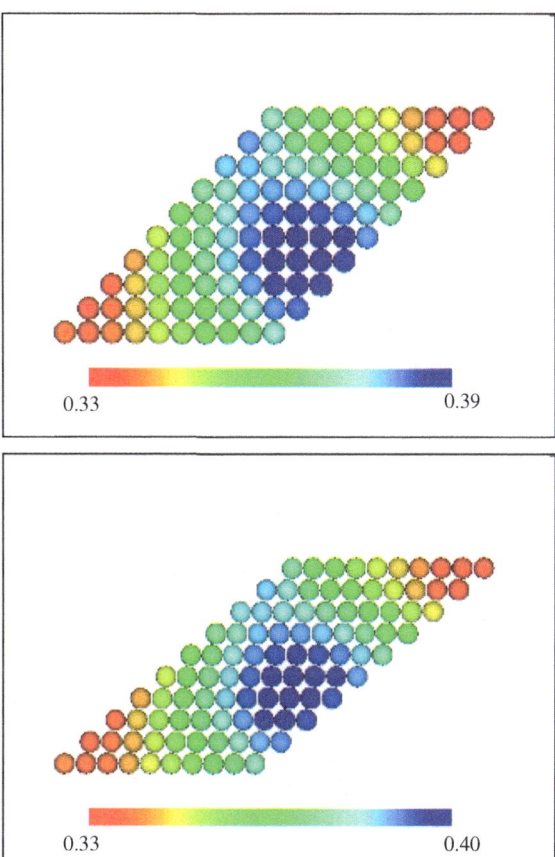

The shear strain increased with constant amplitude. The shear stress of the soil sample gradually increased and then began to stabilize. Since the principal stress ratio was set to M = 1.55 in the SPH program, according to the definition of the Cambridge model, when stress ratio M = 1.55, the soil specimen reaches a critical state. The specimen will fail if the load continues.

The simulated stress–strain relationship and p–q relationship are shown in Figs. 4.37 and 4.38. The CSL of the soil sample is also shown in Fig. 4.38.

To further verify the accuracy of the SPH program with the Cambridge model for the soil non-drained shear test simulation, the Cam-Clay finite element program was used to conduct the same simulation and compare the corresponding results . The computation parameters used in this simulation example are listed in Table 4.8.

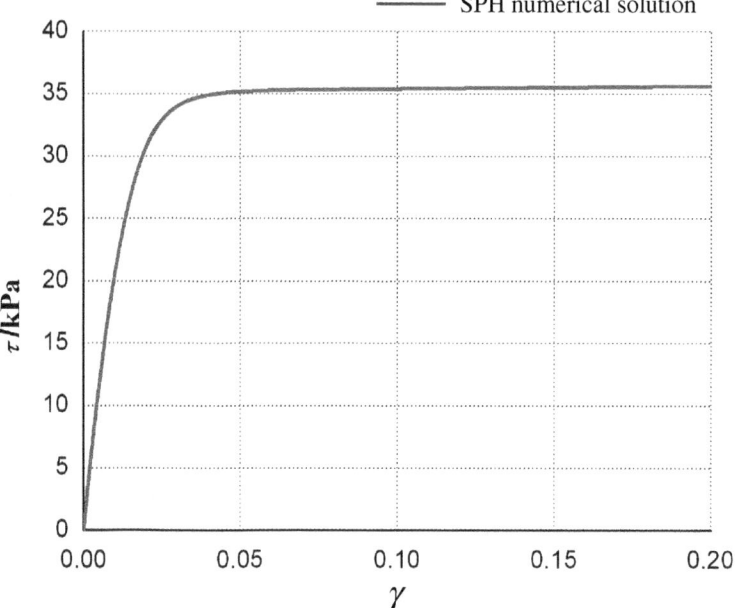

Fig. 4.37 SPH-simulated stress–strain relationship in simple shear condition

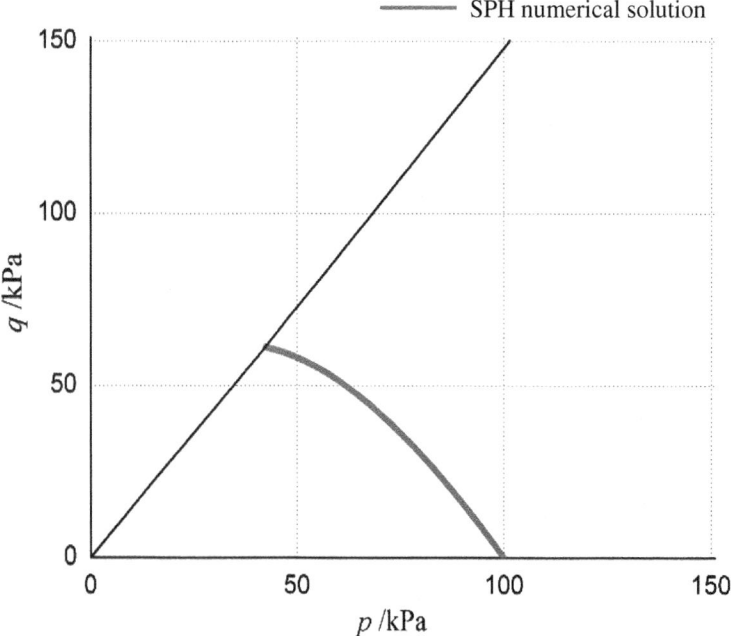

Fig. 4.38 SPH-simulated p–q relationship in simple shear condition

Table 4.8 Parameters used in finite element simulation

Poisson's ratio υ	Compression index λ	Swelling index κ	Initial void ratio e_0	Principal stress ratio R_f	Reference stress p_0 (kPa)	Initial average stress p (kPa)
0.33	0.355	0.0577	1.55	5.12	100	100

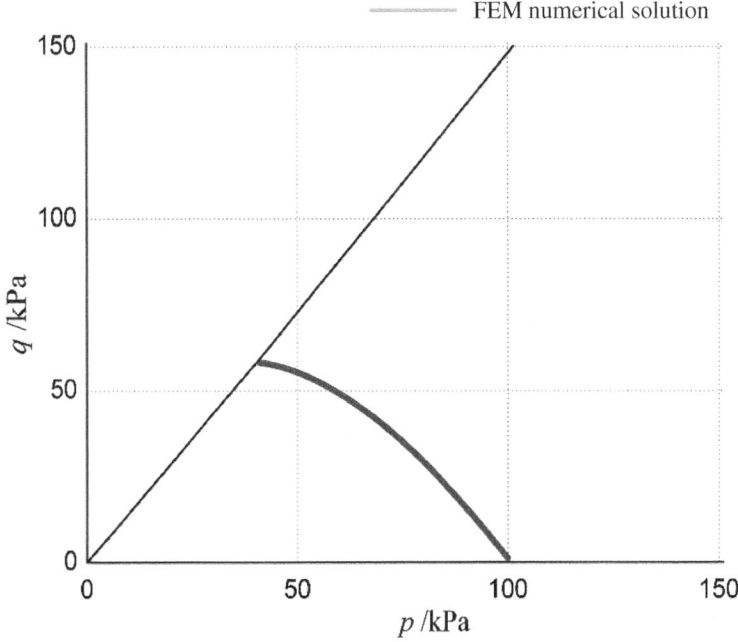

Fig. 4.39 Simulated p–q relationship from FEM program

In the finite element program, the same non-drained strain-controlled condition was used. The termination loading condition of the shear strain was $\varepsilon_{max} = 0.2$ and the loading history was divided into 1,000 time steps. p, q, and τ were output at each time step. The curves of τ–γ and p–q relationships are shown respectively in Figs. 4.39 and 4.40.

Figure 4.41 compares results from the two numerical methods.

From the comparisons in Figs. 4.41 and 4.42, the SPH simulation results for the soil non-drained simple shear test are nearly the same as those from the FEM simulation; their correlation coefficient is 0.92.

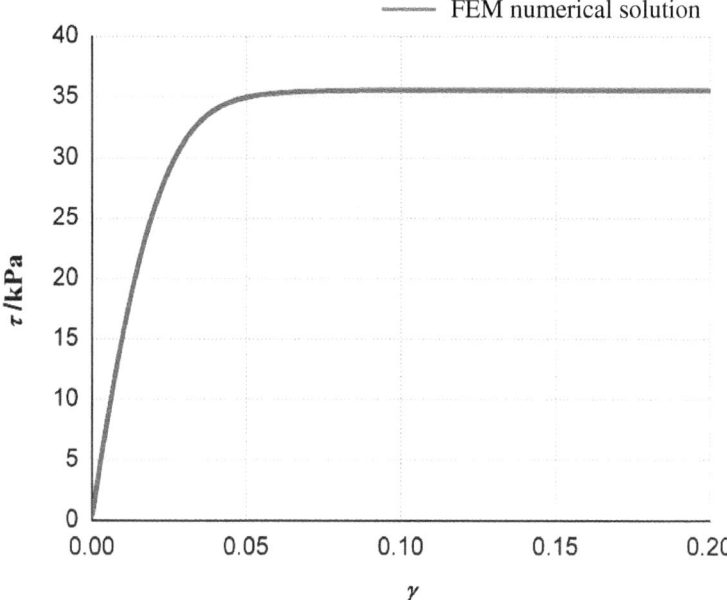

Fig. 4.40 Simulated τ–γ relationship from FEM program

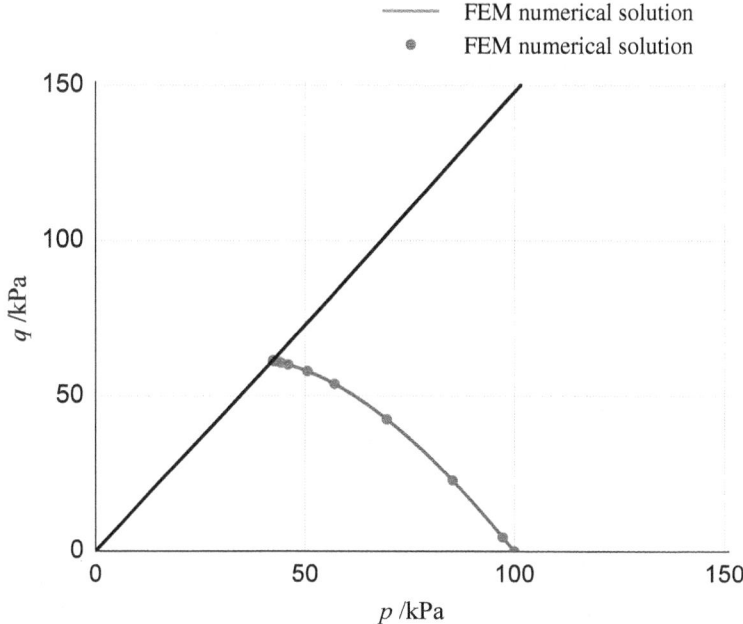

Fig. 4.41 Comparison of p–q relationships from FEM and SPH simulation

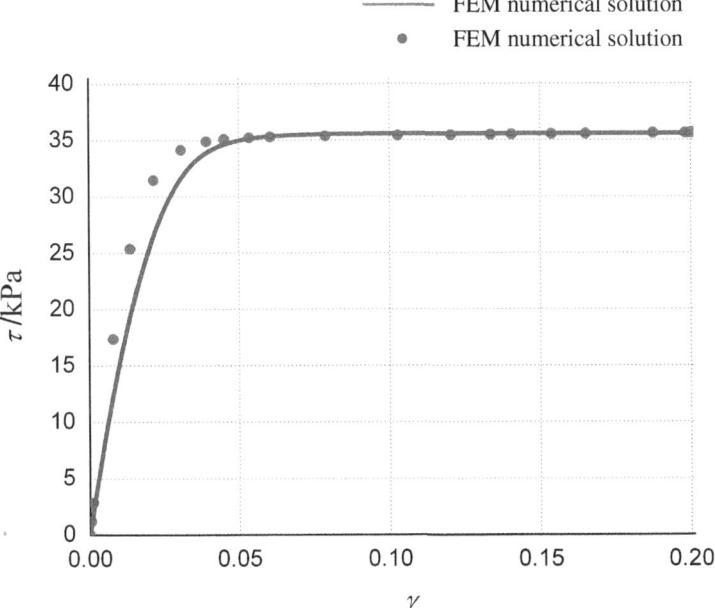

Fig. 4.42 Comparison of τ–γ relationships from FEM and SPH simulation

4.6 Summary

In this chapter, a series of numerical examples were simulated to verify the SPH procedure developed in the Chap. 3, including the dam-break problem, sand flow, slump test for viscous material, simple shear test of elastic material, and soil undrained shear test.

In summary, the conclusions are as follows.

(1) A dam-break model was simulated using the hydrodynamics SPH model. Results were compared with the test results, revealing satisfactory consistency. There was a small error when comparing with the Ritter dam-break theory. However, with the increase of computation time step, the error gradually decreased toward zero.

(2) The Bingham fluid model has been widely applied to numerical analysis of soil material. Therefore, it was applied in combination with the Mohr-Coulomb yield criterion to describe large soil deformation, based on the concept of the equivalent Newtonian viscosity coefficient. The soil flow model test was simulated with good results.

(3) The SPH numerical simulation for the slump test of viscous material was performed. The results were compared with those from the CIP method. The configuration of viscosity material simulated by these two models had good consistency.

(4) The simple shear test for elastic material was simulated. Evolution of the shear strain was simulated, with the maximum strain reaching 2.9. The stress–strain relationship of the elastic material was compared with the analytical solution. Error was less than 1 %, which showed the goodness of fit between the simulation results and theoretical solutions. The validity and reliability of the SPH model were thus confirmed.

(5) The elasto-plastic SPH procedure was used for the soil large-deformation simulation, based on the modified Cambridge model. The soil undrained shear test was simulated and τ–γ and p–q relationships from the SPH method were compared with those from the FEM program, thereby verifying SPH model accuracy.

In summary, a reasonable choice of numerical simulation for different problems can effectively enhance efficiency and accuracy. This chapter detailed several numerical approaches for the SPH method, and some numerical examples were examined to verify the effectiveness of the model. Through numerical analysis of typical examples, the strong adaptability of the SPH method was again demonstrated.

References

Chen, Z. Y., Dai, Z. L., Huang, Y., & Bian, G. Q. (2013). Numerical simulation of large deformation in shear panel dampers using smoothed particle hydrodynamics. *Engineering Structures, 48*, 245–254.

Huang, Y., Dai, Z.L., Zhang, W.J., Chen, Z.Y. (2011). Visual simulation of landslide fluidized movement based on smoothed particle hydrodynamics. *Natural Hazards, 59*(3), 1225–1238.

Huang, Y., Zhang, W. J., Xu, Q., Xie, P., Hao, L. (2012). Run-out analysis of flow-like landslides triggered by the Ms 8.0 2008 Wenchuan earthquake using smoothed particle hydrodynamics. *Landslides, 9*(2), 275–283.

Liu, G.R., Liu, M.B. (2003). Smoothed particle hydrodynamics: a meshfree particle method. World Scientific Publishing Co. Pte. Ltd., Singapore.

Martin, J. C., & Moyce, W. J. (1952). An experimental study of the collapse of liquid columns on a rigid horizontal plane. *Philosophical Transactions of the Royal Society, A255*, 312–325.

Moriguchi, S., Borja, I., Yashima, A., & Sawada, K. (2009). Estimating the impact force generated by granular flow on a rigid obstruction. *Acta Geotech, 4*(1), 57–71.

Moriguchi, S. (2005). *CIP-based numerical analysis for large deformation of geomaterials.* Gifu: Gifu University.

Quecedo, M., Pastor, M., Herreros, M. I., Fernández, J. A., & Zhang, Q. (2005). Comparison of two mathematical models for solving the dam break problem using the FEM method. *Computer Methods in Applied Mechanics and Engineering, 195*(136–38), 3985–5005.

Ritter, A. (1892). Die Fortpflanzung der Wasserwellen. Zeitschrift des Vereins Deutscher Ingenieure, 947–954.

Roscoe, K. H., Schofield, A. N., & Thurairajah, A. (1963). Yielding of clays in states wetter than critical. *Geotechnique, 13*(3), 250–255.

Uzuoka, R., Yashima, A., Kawakami, T., & Konrad, J. M. (1998). Fluid dynamics based prediction of liquefaction induced lateral spreading load. *Computers and Geotechnics, 22*(3–5), 253–282.

Zeng, Y. (2005). *An application of BGK Boltzmann theory to one-dimensional dam-break flow.* Wuhan: Wuhan University.

Chapter 5
SPH Modeling for Flow Slides in Landfills

5.1 Introduction

With economic development and urbanization, the population of and consumption in urban areas increases rapidly, resulting in a growing amount of municipal solid waste (MSW) produced by people, offices, and small industries. This consists of components such as household, hospital, and construction waste. According to a report of the Ministry of Environmental Protection (2011), the production of MSW in China as of the end of 2010 was more than 2.4 billion tons per year, and this increased at an annual rate of 18 % in tandem with economic development.

At present, there are four primary methods for the treatment and disposal of MSW: landfill, incineration, composting, and recycling (Nie et al. 2000). Of these methods, though recycling and reuse of MSW have received increased attention in many countries, landfill is still the main disposal method of dumped MSW, and will remain so for a long time because of its great advantage in cost-benefit analyses (Blight and Fourie 2005). Research has shown that about 79 % of MSW in China enters landfills (Wang and Nie 2001). In the landfill environment, dumped MSW can convert to liquid or even gas and increase water and gas pressure, thereby changing the skeletal structure and mechanical properties of the landfill. Under the influence of external factors such as leachate recharging, heavy rainfall, and earthquakes, landfill slopes can easily become unstable. Many even evolve into large-scale flow slides of high velocity and long runout.

Six large-scale failures in MSW dumps and landfills have been described in a review by Blight (2008), all of which were flow slides with great volumes and long runout distances. For instance, the first reported waste slide was in Sarajevo, Yugoslavia in the 1970s, which involved 20,000 m^3 of refuse flowing more than 1 km down a gently sloping hillside. Two bridges and five houses were destroyed (Blight 2008). A landfill in Bandung, Indonesia failed after 3 days of heavy rainfall, flowing nearly 1 km down a canyon. In addition, other landfill flow slides

© Springer-Verlag Berlin Heidelberg 2014
Y. Huang et al., *Geo-disaster Modeling and Analysis: An SPH-based Approach*,
Springer Natural Hazards, DOI 10.1007/978-3-662-44211-1_5

have been recorded, such as slope failure in an MSW landfill near Cincinnati in 1996 (Chugh et al. 2007) and the Kettleman Hills landfill slope failure in 1988 (Mitchell et al. 1990). These flow failures killed more than 500 people, filled many streambeds, destroyed numerous bridges and houses, and caused a leak of dumped MSW and leachate over a large area. Such toxic MSW and leachate can seriously pollute soil, plants, surface water, groundwater, and air in surrounding areas, resulting in substantial environmental cleanup and repair costs (Liu et al. 2007). It follows that landfill flow slides are associated with extreme danger and serious economic loss. There is therefore an urgent need to obtain a clearer understanding of the dynamic behavior of waste during its fluidized movement, which will ultimately play an indispensable role in the prevention and assessment of waste slide hazards.

Numerical simulations have been widely used as efficient tools in the analysis of landfill slope stability. Available three-dimensional analysis methods have been reviewed, and a novel model has been proposed based on the limit equilibrium method. This properly considered arbitrary slip surface geometry, sliding kinematics, and the interslice force to analyze slope stability of the Kettleman Hills landfill, and correctly predicted both pre-slide and post-slide configurations (Chang 2002, 2005). Finite difference method (FDM) computer programs FLAC and FLAC3D were used to simulate the observed translational character of a landfill slide failure near Cincinnati (Chugh et al. 2007). An extended environmental multimedia modeling system (EEMMS) was developed to characterize the dynamics involved in typical environmental multimedia problems, using both the finite element method (FEM) and FDM (Chen and Yuan 2009). The TMC-Slope FEM program was developed on the basis of the limited balance method to analyze landfill stability and determine the failure surface (Liu et al. 2010). A hybrid method for quasi-three-dimensional slope stability analysis based on finite element stress analysis was applied in a case study of an MSW landfill in northeastern Spain (Yu and Batlle 2011). All these methods are based on a framework of solid mechanics and focus on stability analysis, and have achieved some promising results. However, mesh-based methods such as FEM and FDM can suffer from numerical difficulties (e.g., severe mesh winding, twisting, and distortion), particularly for flow slides with extremely large deformations. Consequently, remeshing is needed during the course of the solution, which generally increases the complexity of computer programs and reduces their calculation precision (Chen and Beraun 2000). As a result, the methods are unable to either simulate motion of the flow front or estimate runout of the waste flow slide. Therefore, a more advanced numerical method suitable for large deformation analysis is urgently needed to simulate flow failure of MSW landfills.

Over the past few years, meshless methods have been a major research focus, aiming at the next generation of effective computational tools for complicated problems (Liu and Liu 2010). Of these methods, Smoothed Particle Hydrodynamics (SPH) is a recently developed, mesh-free numerical method that originated in an astrophysics application (Lucy 1977; Gingold and Monaghan 1977). As a non-mesh-based technique, the main advantage of SPH is that it bypasses the need for a numerical grid and avoids severe mesh distortions caused by large deformation (Wang et al. 2005). Because of this advantage, SPH has been widely and successfully applied in the field of fluid dynamics in the past decade (Fang et al. 2006; Xiong et al. 2006). Recently, applications of this method have gradually developed and been applied to large deformation analysis within various disciplines, such as metal manufacturing (Cleary et al. 2006), strong shocks and high-velocity collisions (Sigalotti et al. 2009; Johnson et al. 1993), rock caving (Karekal et al. 2011), and buried structures (Wang et al. 2005; Lu et al. 2005). Our group introduced SPH combined with a Bingham constitutive model to simulate fluidized movement of flow-like landslides (Huang et al. 2011a, 2012) and lateral spread of liquefied soils (Huang et al. 2011b), with some positive results.

Though the SPH method has been widely applied in many fields of engineering, SPH simulations of flow failure in MSW landfills are rare in the literature. Because of the lack of published studies in this field, SPH is applied in this chapter as a means for simulating the dynamic behavior of dumped MSW. The landfill configuration after failure is predicted and runout of the waste slide is estimated, both of which are extremely important in hazard prevention and risk evaluation in the landfill area.

5.2 Flow Slide in Sarajevo Landfill

Flow failure of the MSW landfill in Sarajevo (then a city with population of 3,50,000) occurred in December 1977. This was apparently the first flow slide of MSW recorded in the technical literature. The main components of MSW in this landfill according to Blight (2008) were food and garden waste, cardboard, paper, plastic, and other loose and uncompacted waste. The loose MSW was stacked on a steep hillside at an angle of about 20°, forming a landfill around 50 m in height. Around 2,00,000 m^3 of waste liquefied and then flowed rapidly down the hillside. It traveled more than 1 km, rushed onto the valley floor, and buried two bridges and five houses, filling two streambeds with waste (Blight 2008). Figure 5.1 shows a longitudinal section of this MSW flow slide. The failure surface and configurations before and after the failure are shown.

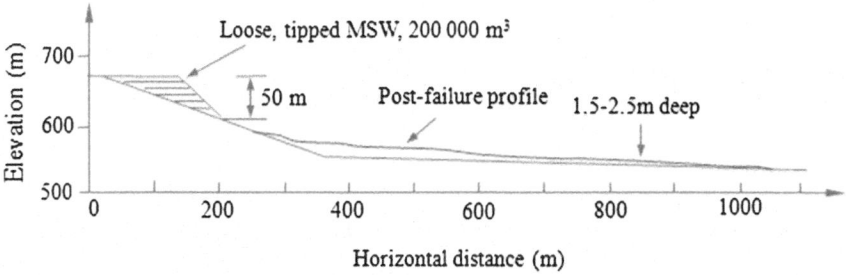

Fig. 5.1 Section of MSW flow slide at Sarajevo, Yugoslavia in 1977 (based on Blight 2008)

Using the SPH modeling technique, flow failure simulation of the Sarajevo landfill was conducted to study the dynamic behavior of MSW during its movement and analyze the final runout. The resulting numerical model is shown in Fig. 5.2a. There were 2,619 particles in total, 1,653 for the MSW and 966 for the boundary, with 2 m space between particles. Just as in the actual situation, the

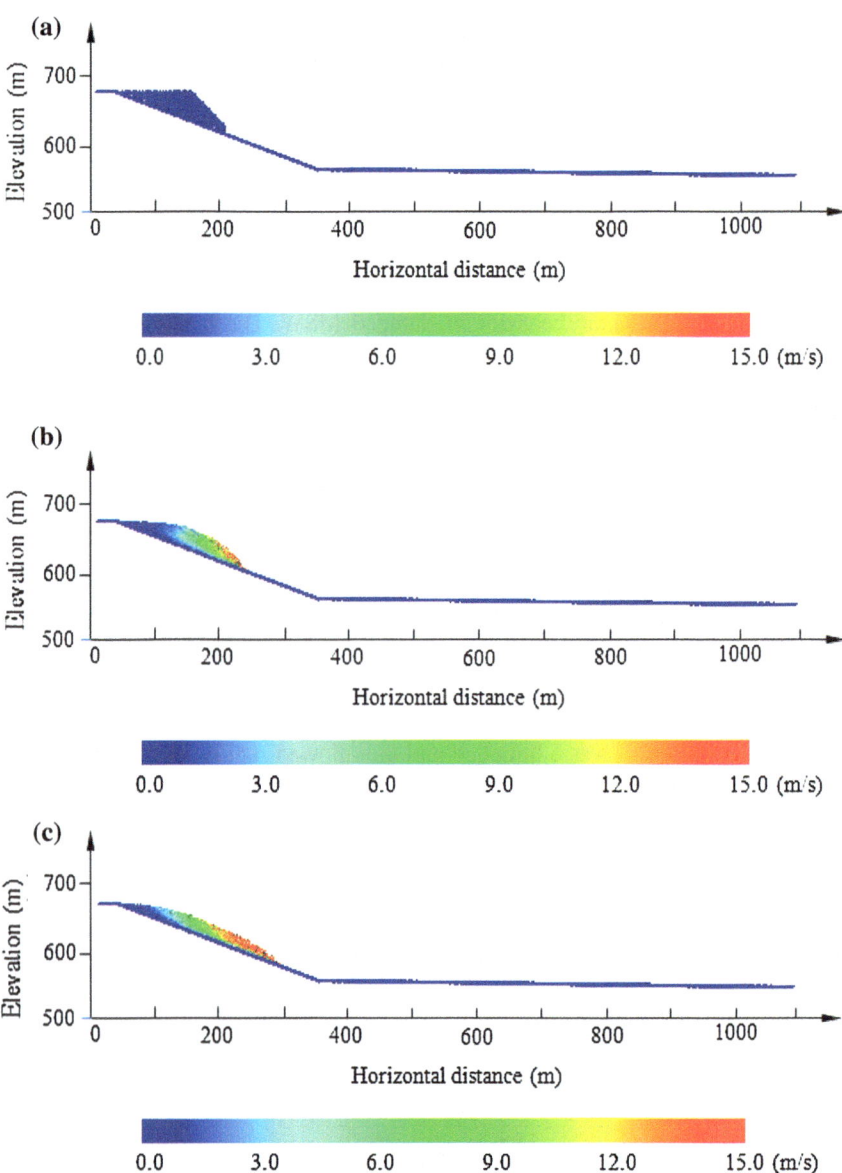

Fig. 5.2 Simulated flow process of Sarajevo landfill. **a** $t = 0.0$ s, **b** $t = 3.0$ s, **c** $t = 6.0$ s, **d** $t = 12.0$ s, **e** $t = 24.0$ s, **f** $t = 150.0$ s, **g** $t = 300.0$ s, **h** $t = 600.0$ s

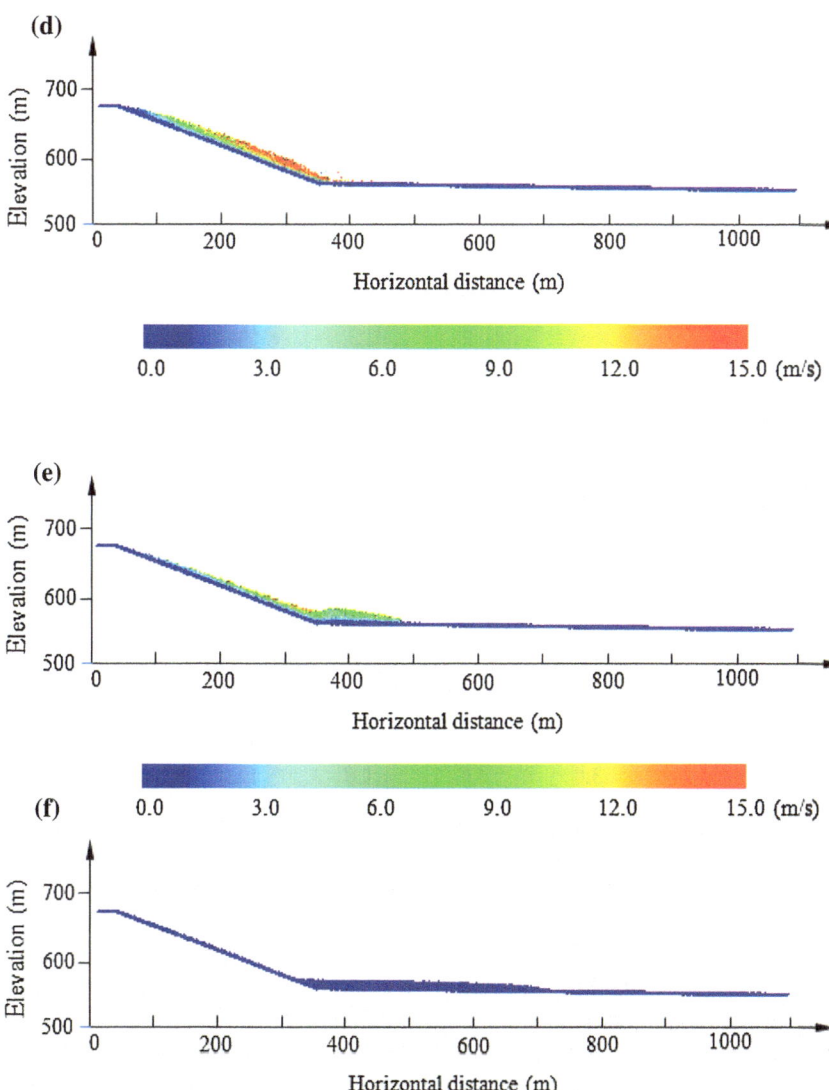

Fig. 5.2   (continued)

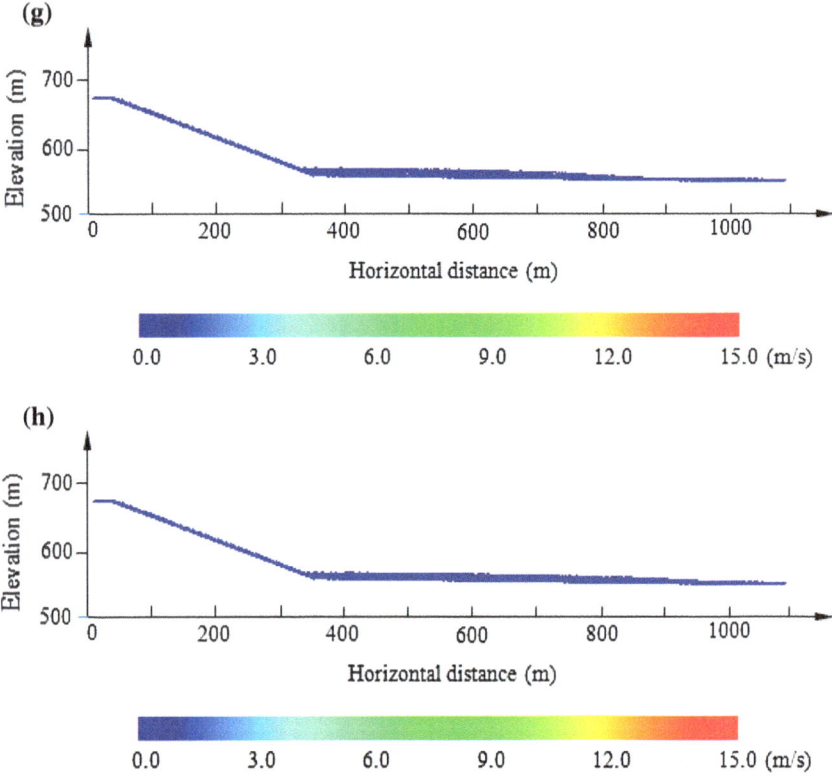

Fig. 5.2 (continued)

MSW particles can be deformed both horizontally and vertically, with gravitation applied only in the vertical direction.

The selection of suitable constitutive models is an important part of the numerical simulation. Since the MSW composition is extremely complex, it is difficult to find a suitable model to describe its stress–strain character. In the present study, a series of laboratory tests were performed to obtain the peak, ultimate, and residual shear stress, and corresponding shear deformation of the filled waste and the interface between geosynthetic layers. These tests have included those of conventional direct shear (Mitchell et al. 1990; Bergadoa et al. 2006; Bacas et al. 2011), large-scale direct shear (Cho et al. 2011; Zekkos et al. 2010), and torsional ring shear (Stark and Poepple 1994; Eid 2011). The test results all provided fundamental and reliable strength parameters for numerical simulation. However, their constitutive relationships were all in the framework of solid mechanics and only suitable for description of the constitutive behavior over a certain range of strain. These relationships are invalid for complicated problems associated with extremely large deformation such as landfill flow slides, in which the strain rate is much greater than 100 %. In such a situation, large deformation results in negligible stiffness of the MSW, and the ground behaves intrinsically as a fluid. Therefore, constitutive models based on solid mechanics are unsuitable for describing the constitutive relationship of a flow-like MSW.

Liquefaction of MSW has been identified as one of the key reasons for flow failure in landfills. For example, in the failure of landfills at Sarajevo, Umraniye-Hekimbashi, Payatas, and Bandung (Blight 2008), MSW showed a mobility indicative of static liquefaction. Therefore, an MSW flow slide may be regarded as a multiphase flow of liquefied geomaterials. Uzuoka et al. (1998) treated liquefied soil as a Bingham fluid and carried out several numerical analyses supported by experimental results. Hadush et al. (2000) summarized the relationship between viscosity and shear strain rate based on three measurement methods, suggesting that the relationship between this rate and shear stress in highly deformed soil materials is in good agreement with the Bingham flow model. Given a lack of similar research on constitutive models for fluidized MSW, we use the Bingham model in this chapter to describe the fluidization characteristics of MSW in SPH simulation.

Simulation parameters are listed in Table 5.1. The shear strength characteristics of MSW, cohesion c, and inner friction angle φ have been studied through a series of laboratory tests and in situ landfill investigations in several countries and regions, including Meruelo in Spain (Sanchez-Alciturri et al. 1993), Liossia in Greece (Coumoulos et al. 1995), Hangzhou and Suzhou in China (Chen et al. 2003; Zhan et al. 2008), Illinois in the United States (Reddy et al. 2009), and Salvador in Brazil (Machado et al. 2010). Unfortunately, no shear strength characterization of MSW at the Sarajevo landfill was found in the technical literature. Therefore, c and φ used in the SPH simulation here were given average values of the experimental results in the aforementioned research. The viscosity coefficient η has been confirmed to be insensitive to the simulated velocity of liquefied soil (Uzuoka et al. 1998). Owing to a lack of relevant experimental data of this coefficient for on-site MSW, we used the viscosity coefficient 1.0 Pa·s for simulations. This value is the same as that of Uzuoka et al. (1998) in their analysis of liquefaction-induced lateral spreading.

Figure 5.2 shows that SPH could reproduce the entire flow process of MSW in the Sarajevo landfill and show the evolution of the final slide shape. The color legend in this figure represents the particle velocity in m/s, with values increasing gradually from blue to red. One can see from the figure that: (a) maximum velocity during the flow slide was around 15 m/s; (b) the runout was about 800 m from the toe of the landfill; and (c) average thickness of the MSW sediment was about 4 m. To reveal the quality of SPH analysis for the landfill flow failure, a comparison of SPH-simulated geometry and landfill configuration surveyed after

Table 5.1 Parameters used in SPH simulation of Sarajevo landfill

Density	ρ (kg/m^3)	1050
Cohesion	c (kPa)	16
Angle of internal friction	φ (°)	22
Viscosity coefficient	η (Pa·s)	1.0
Acceleration of gravity	g (m/s^2)	9.8
Unit time step	Δt (s)	0.03
Time step	n	20000

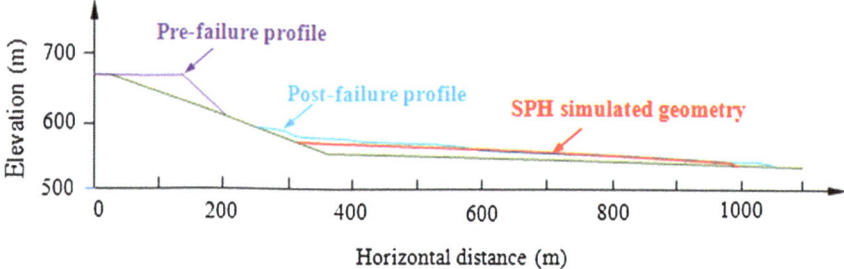

Fig. 5.3 Pre- and post-failure profiles and comparison of SPH simulation with survey data for Sarajevo landfill (Huang et al. 2013)

the disaster is shown in Fig. 5.3. It is obvious that despite slight deviation at the toe of the hillside, the runout, coverage of dumped waste, and sediment thickness were very similar to those surveyed on-site.

5.3 Flow Slide in Bandung Landfill

Bandung is the capital of the West Java province in Indonesia, with a population of 6 million. A landfill in Bandung failed on 21 February 2005, accompanied by a sound like rolling thunder. The waste liquefied after 3 days of heavy rainfall, resulting in about 2.7×10^6 m^3 of waste flowing down the hillside, covering a 200–250 m wide strip 900 m long (Koelsch et al. 2005). The valley floor below the landfill was covered by waste. After the slide, the bodies of 147 shack-dwellers living on the dump were recovered, but the number of bodies remaining in the slide is unknown. Figure 5.4 shows a longitudinal section of this landfill before and after the failure, from which we see that the elevation difference between the front and back edges of the slide was about 65 m. Angles of the tripping face and failure surface were about 40 and 60°, respectively.

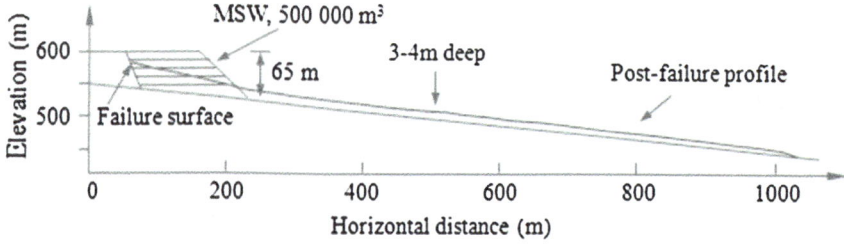

Fig. 5.4 Section of MSW flow slide at Bandung, Indonesia in 2005 (based on Blight 2008)

As for the Sarajevo event, numerical simulation of the entire flow process of the Bandung landfill was conducted using the SPH model. Figure 5.5 shows that the numerical model had 2,188 particles in total, 1,290 for the MSW and 898 for the boundary, with 3 m particle spacing. Referring to Koelsch et al. (2005), density ρ, c, and φ of MSW in this landfill were taken as 1,100 kg/m^3, 10 kPa, and 20°, respectively (Table 5.2).

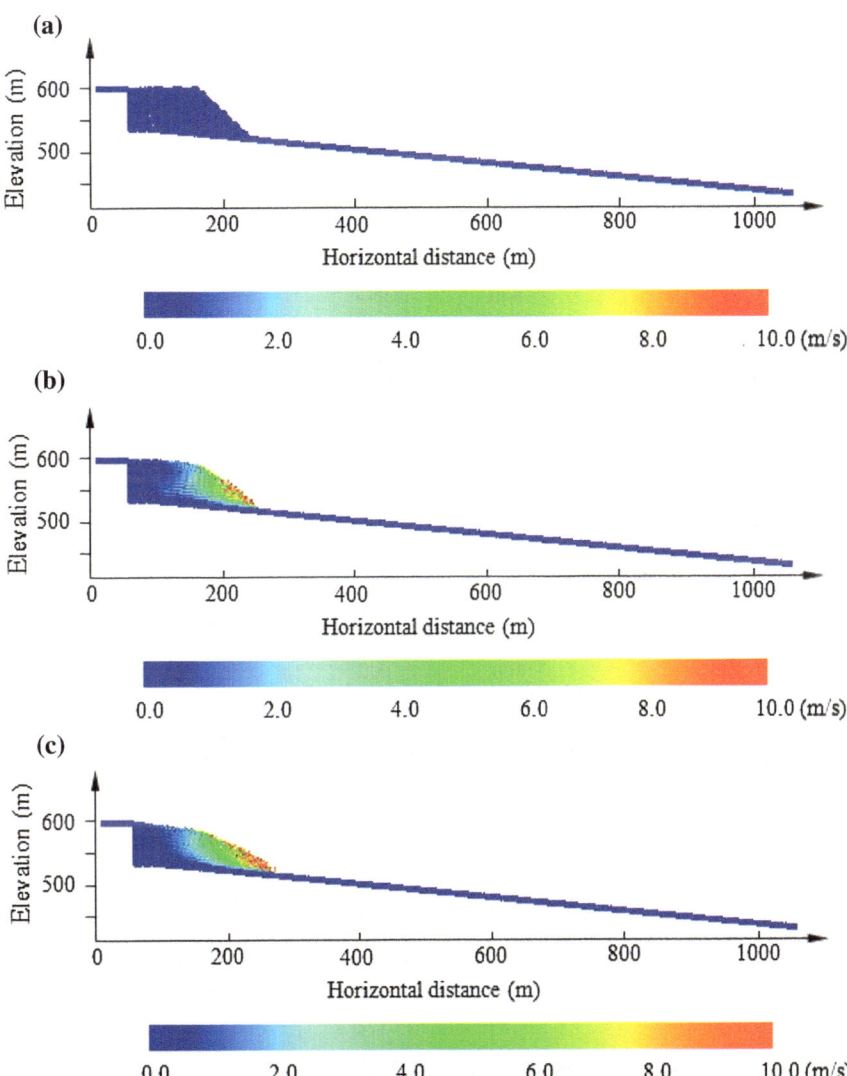

Fig. 5.5 Simulated flow process of Bandung landfill. **a** $t = 0.0$ s, **b** $t = 7.5$ s, **c** $t = 15.0$ s, **d** $t = 30.0$ s, **e** $t = 60.0$ s, **f** $t = 120.0$ s, **g** $t = 240.0$ s, **h** $t = 480.0$ s, **i** $t = 900.0$ s

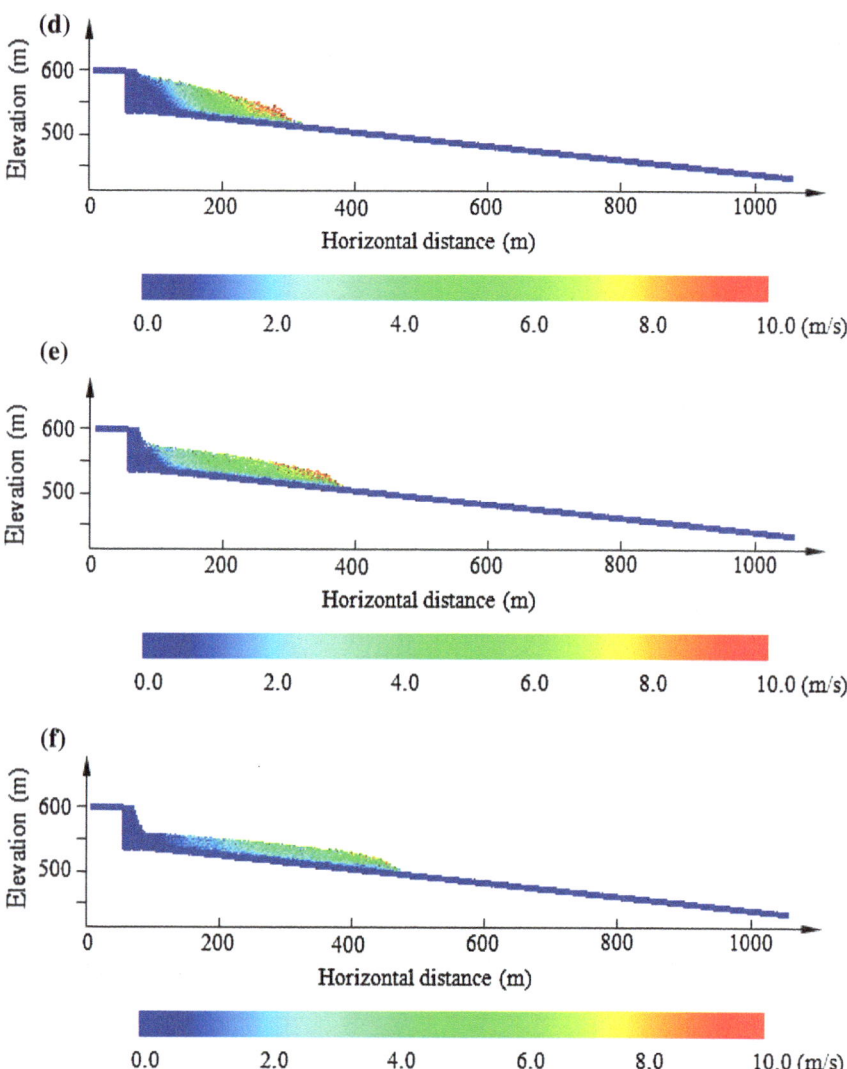

Fig. 5.5 (continued)

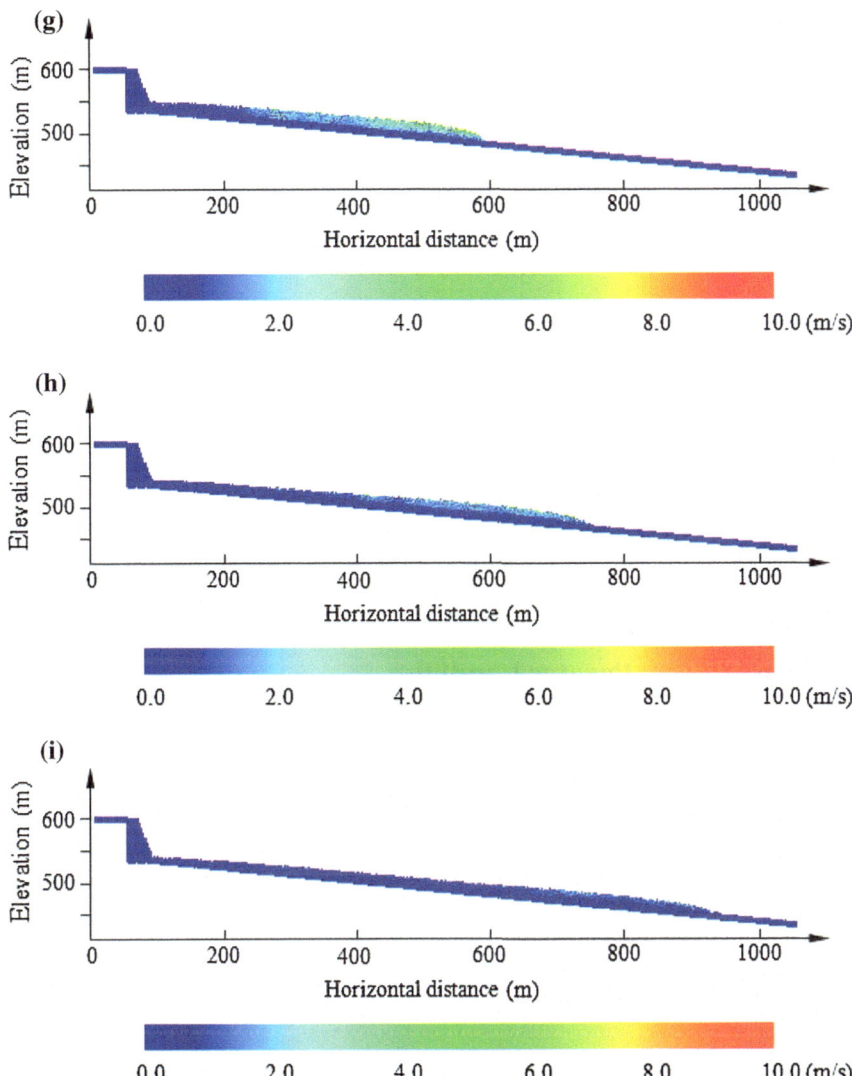

Fig. 5.5 (continued)

Table 5.2 Parameters used in SPH simulation of Bandung landfill			
Density	ρ (kg/m^3)	1100
Cohesion	c (kPa)	10
Angle of internal friction	φ (°)	20
Viscosity coefficient	η (Pa·s)	1.0
Acceleration of gravity	g (m/s^2)	9.8
Unit time step	Δt (s)	0.03
Time step	n	30000

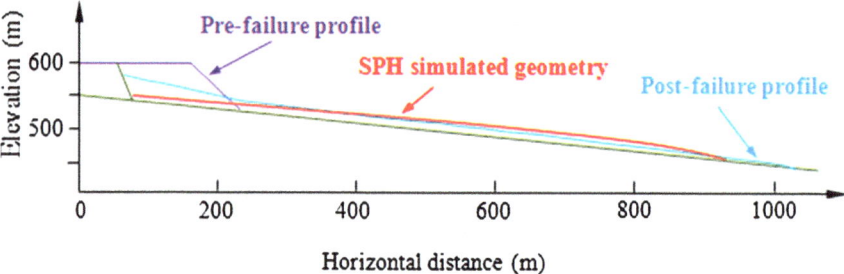

Fig. 5.6 Pre- and post-failure profiles and comparison of SPH simulation with survey data for Bandung landfill (Huang et al. 2013)

Figure 5.5 portrays the motion of the waste slide in Bandung landfill. From the numerical results, maximum velocity during the flow slide was around 10 m/s and the runout exceeded 700 m. The simulated post-failure profile was compared with the surveyed configuration (Fig. 5.6). Despite slight deviation in the modeled slope configuration, the runout simulation was very similar to the result measured at the site.

5.4 Flow Slide in Payatas Landfill

Another typical flow slide at an MSW landfill occurred at Payatas in the Philippines. In 2000, MSW at this landfill failed under the action of rainfall. Nearly 15,000 m^3 of waste flowed downhill. Although the failed MSW only flowed ~20 m, it destroyed all buildings in the way and killed 100 people. A section of the MSW flow slide at Payatas landfill is shown in Fig. 5.7.

SPH simulation of the flow slide at Payatas was performed, using the parameters listed in Table 5.3.

Fig. 5.7 Section of MSW flow slide at Payatas landfill

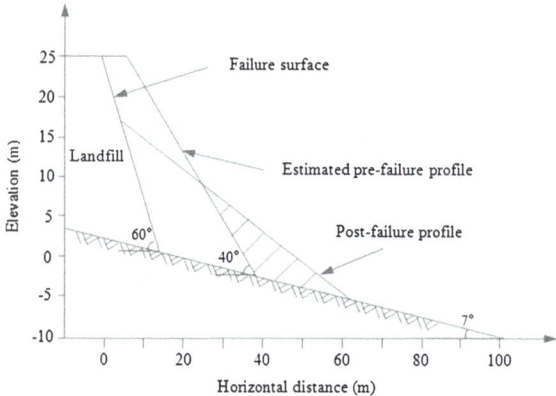

Table 5.3 Parameters used in SPH simulation of Payatas landfill

Density	ρ (kg/m^3)	1100
Cohesion	c (kPa)	10
Angle of internal friction	φ ($^\circ$)	20
Viscosity coefficient	η (Pa·s)	1.0
Acceleration of gravity	g (m/s^2)	9.8
Unit time step	Δt (s)	0.005
Time step	n	3000

Figure 5.8 shows the final configuration of the Payatas landfill after the flow slide, with the color bar again representing particle velocity. The final configuration was compared with the one measured on-site. This showed simulated and measured results in good agreement.

5.5 Analysis of Simulation Results

The three typical landfill flow failures modeled and discussed above verify the accuracy and reliability of the SPH model, and extend its application to flow slide analyses in landfill settings. Comparisons of the SPH results with surveyed waste slide configurations for these two landfills show that SPH-simulated geometries were very similar to surveyed ones. This suggests that SPH simulations can accurately reproduce full flow processes of landfill flow slides.

In addition to slope configurations, other fundamental dynamic behaviors can be derived from SPH analysis, including the flow velocity during the motion. Based on an understanding of these dynamic behaviors, hazard assessments and risk evaluations can be put into practice. This will ultimately help decision makers make appropriate recommendations to avoid potential disasters. However, with the exception of slope configurations surveyed post-disaster, there are no real-time data (e.g., slide velocity information) from these waste slides. Therefore, there is still a lack of information on the entire flow process. As a result, the simulations herein focused on comparisons of hillside profiles before and after failure.

Although further calibration and validation are required for flow-slide simulation of MSW landfills, from the comparison of actual and modeled runout configurations, SPH can be an effective tool for investigating the flow characteristics of waste slides, and can capture essential characterization parameters such as runout and velocity.

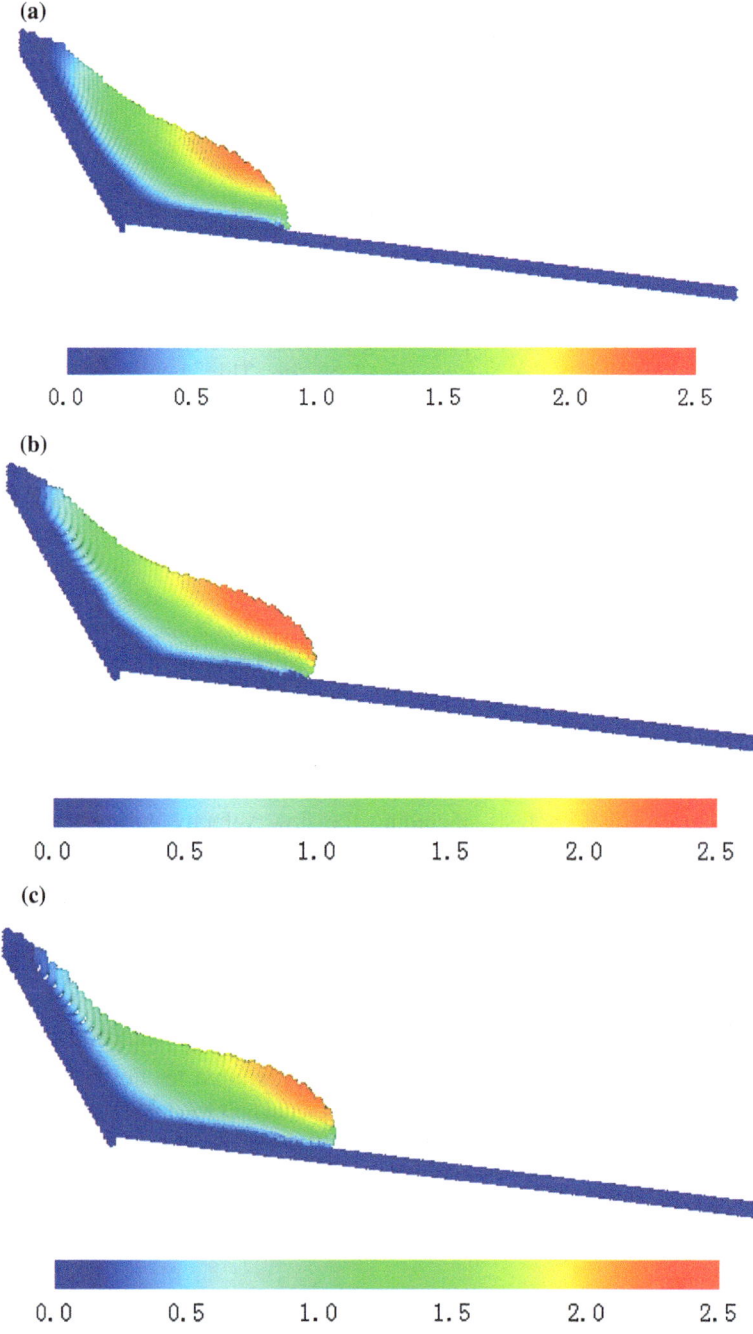

Fig. 5.8 Simulated flow process of Payatas landfill. **a** $t = 2.5$ s, **b** $t = 5$ s, **c** $t = 7.5$ s, **d** $t = 10$ s, **e** $t = 12.5$ s, **f** $t = 15$ s

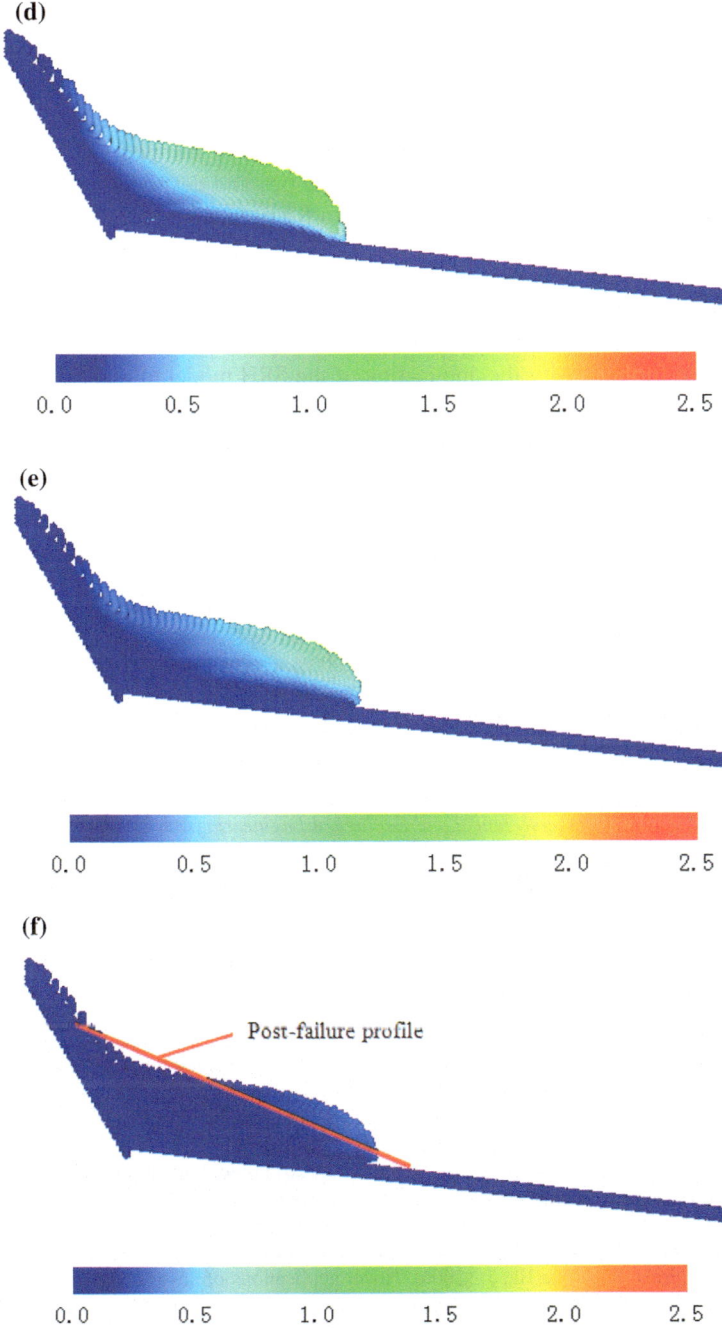

Fig. 5.8   (continued)

5.6 Summary

Catastrophic flow slides have occurred with increasing frequency in MSW landfills, even in those that were carefully controlled and well engineered. In addition to the environmental contamination caused by the diffusion of dumped waste and toxic leachate, there have been large numbers of fatalities and destruction of infrastructure. Numerical simulations are very beneficial for deepening our understanding of flow characteristics in such waste slides.

The failures at the Sarajevo, Bandung, and Payatas landfills are typical examples of flow slides in MSW dumps that caused significant damage and casualties. SPH simulations of these landfill failures, for which field data were available, were conducted to further verify the application of the SPH model to real flow slides in MSW landfills. The propagation of flow failures in these landfills was represented. This chapter presented preliminary results of runout analysis of landfill failures.

References

Bacas, B. M., Konietzky, H., Berini, J. C., & Sagaseta, C. (2011). A new constitutive model for textured geomembrane/geotextile interfaces. *Geotextiles and Geomembranes, 29*, 137–148.

Bergadoa, D. T., Ramanab, G. V., & Varun, H. I. S. (2006). Evaluation of interface shear strength of composite liner system and stability analysis for a landfill lining system in Thailand. *Geotextiles and Geomembranes, 24*, 371–393.

Blight, G. E. (2008). Slope failures in municipal solid waste dumps and landfills: A review. *Waste Management and Research, 26*(5), 448–463.

Blight, G. E., & Fourie, A. B. (2005). Catastrophe revisited—disastrous flow failures of mine and municipal solid waste. *Geotechnical and Geological Engineering, 23*, 219–248.

Chang, M. (2002). A 3D slope stability analysis method assuming parallel lines of intersection and differential straining of block contacts. *Canadian Geotechical Journal, 39*(4), 799–811.

Chang, M. (2005). Three-dimensional stability analysis of the Kettleman Hills landfill slope failure based on observed sliding-block mechanism. *Computers and Geotechnics, 32*, 587–599.

Chen, J. K., & Beraun, J. E. (2000). A generalized smoothed particle hydrodynamics method for nonlinear dynamic problems. *Computer Method in Applied Mechanics and Engineering, 190*(1–2), 225–239.

Chen, Y. M., Luo, C. Y., & Ke, H. (2003). Geotechnical properties of municipal solid waste in China. In *Proceedings of the 12th Asian Regional Conference on Soil Mechanics and Geotechnical Engineering* (Vol. 1–2, pp. 365–368).

Chen, Z., & Yuan, J. (2009). An extended environmental multimedia modeling system (EEMMS) for landfill case studies. *Environmental Forensics, 10*, 336–346.

Cho, Y. M., Koa, J. H., Chi, L. Q., & Townsend, T. G. (2011). Food waste impact on municipal solid waste angle of internal friction. *Waste Management, 31*(1), 26–32.

Chugh, A. K., Stark, T. D., & DeJong, K. A. (2007). Reanalysis of a municipal landfill slope failure near Cincinnati, Ohio, USA. *Canadian Geotechnical Journal, 44*(1), 33–53.

Cleary, P. W., Prakash, M., & Ha, J. (2006). Novel applications of smoothed particle hydrodynamics (SPH) in metal forming. *Journal of Materials Processing Technology, 177*(1–3), 41–48.

Coumoulos, D. G., Koryalos, T. P., Metaxas, J. L., & Gioka, D. A. (1995). Geotechnical investigation at the main landfill of Athens. In: *Proceedings Sardinia 95, 5th International Landfill Symposium*, (pp. 885–895). Cagliari, Italy.

Eid, H. T. (2011). Shear strength of geosynthetic composite systems for design of landfill liner and cover slopes. *Geotextiles and Geomembranes, 29*(3), 335–344.

Fang, J. N., Owens, R. G., Tacher, L., & Parriaux, A. (2006). A numerical study of the SPH method for simulating transient viscoelastic free surface flows. *Journal of Non-Newtonian Fluid Mechanics, 139*(1–2), 68–84.

Gingold, R. A., & Monaghan, J. J. (1977). Smoothed particle hydrodynamics: Theory and application to non-spherial stars. *Monthly Notices of the Royal Astronomical, 181*, 375–389.

Huang, Y., Dai, Z. L., Zhang, W. J., & Chen, Z. Y. (2011a). Visual simulation of landslide fluidized movement based on smoothed particle hydrodynamics. *Natural Hazards, 59*(3), 1225–1238.

Huang, Y., Zhang, W. J., Mao, W. W., & Jin, C. (2011b). Flow analysis of liquefied soils based on smoothed particle hydrodynamics. *Natural Hazards, 59*(3), 1547–1560.

Huang, Y., Dai, Z. L., Zhang, W. J., & Huang, M. S. (2013). SPH-based numerical simulations of flow slides in municipal solid waste landfills. *Waste Management and Research, 31*(3), 256–264.

Huang, Y., Zhang, W. J., Xu, Q., & Xie, P. (2012). Run-out analysis of flow-like landslides triggered by the Ms 8.0 2008 Wenchuan earthquake using smoothed particle hydrodynamics. *Landslides 9*(2), 275–283.

Hadush, S., Yashima, A., & Uzuoka, R. (2000). Importance of viscous fluid characteristics in liquefaction induced lateral spreading analysis. *Computers and Geotechnics, 27*(3), 199–224.

Johnson, G. R., Petersen, E. H., & Stryk, R. A. (1993). Incorporation of an SPH option into the EPIC code for a wide range of high velocity impact computations. *International Journal of Impact Engineering, 14*(1–4), 385–394.

Karekal, S., Das, R., Mosse, L., & Cleary, P. W. (2011). Application of a mesh-free continuum method for simulation of rock caving processes. *International Journal of Rock Mechanics and Mining Sciences, 48*(5), 703–711.

Koelsch, F., Fricke, K., Mahler, C., & Damanhuri, E. (2005). Stability of landfills-the Bandung dumpsite disaster. In *Proceedings of the 10th International Waste Management and Landfill Symposium*. Sardinia, Cagliari, Italy.

Liu, C. L., Zhang, Y., Zhang, F., et al. (2007). Assessing pollutions of soil and plant by municipal waste dump. *Environmental Geology, 52*(4), 641–651.

Liu, M. B., & Liu, G. R. (2010). Smoothed particle hydrodynamics (SPH): An overview and recent developments. *Archives of Computational methods in Engineering, 17*(1), 25–76.

Liu, X. L., Si, W. J., Zhu, C. Y., & et al. (2010). Analysis on stability of liner system located in slope of municipal solid waste landfill. In *Proceeding of 7th International Symposium on Safety Science and Technology (ISSST)*. Hangzhou, China.

Lu, Y., Wang, Z. Q., & Chong, K. R. (2005). A comparative study of buried structure in soil subjected to blast load using 2D and 3D numerical simulations. *Soil Dynamics and Earthquake Engineering, 25*(4), 275–288.

Lucy, L. B. (1977). A numerical approach to the testing of the fission hypothesis. *Astronomical Journal, 82*(12), 1013–1024.

Machado, S. L., Karimpour-Fard, M., Shariatmadari, N., et al. (2010). Evaluation of the geotechnical properties of MSW in two Brazilian landfills. *Waste Management, 30*(12), 2579–2591.

Ministry of Environmental Pretection of the People's Republic of China (2011). The *communique on china's environmental conditions*. Retrieved March 6, 2011, from http://jcs.mep.gov .cn/hjzl/zkgb/2010zkgb/201106/t20110602_211569.htm (in Chinese).

Mitchell, J. K., Seed, R. B., & Seed, H. B. (1990). Kettleman Hills waste landfill slope failure. I. liner-system properties. *Journal of geotechnical engineering, 116*(4), 647–668.

Nie, Y. F., Niu, D. J., & Bai, Q. Z. (2000). The management of municipal solid waste in China. *Journal of Environmental Science and Health Part A-Toxic/Hazardous Substances and Environmental Engineering, 35*(10), 1973–1980.

Reddy, K. R., Hettiarachchi, H., Parakalla, N. S., et al. (2009). Geotechnical properties of fresh municipal solid waste at Orchard Hills landfill USA. *Waste Management, 29*(2), 952–959.

Sanchez-Alciturri, J. M., Palma, J., Sagaseta. C., & et al. (1993). Mechanical properties of wastes in a sanitary landfill. In *Proceedings of the Green '93, Waste Disposal by Landfill* (pp. 357–363). Bolton, UK.

Sigalotti, L. D., Lopez, H., & Trujillo, L. (2009). An adaptive SPH method for strong shocks. *Journal of Computational Physics, 228*(16), 5888–5907.

Stark, T. D., & Poepple, A. R. (1994). Landfill liner interface strengths from torsional-ring-shear tests. *Journal of Geotechnical Engineering, ASCE, 120*(3), 597–615.

Uzuoka, R., Yashima, A., Kawakami, T., et al. (1998). Fluid dynamics based prediction of liquefaction induced lateral spreading. *Computers and Geotechnics, 22*(3/4), 243–282.

Wang, H. T., & Nie, Y. F. (2001). Municipal solid waste characteristics and management in China. *Journal of the Air and Waste Management Association, 51*, 250–263.

Wang, Z. Q., Lu, Y., Hao, H., & Chong, K. (2005). A full coupled numerical analysis approach for buried structures subjected to subsurface blast. *Computers and Structures, 83*(4–5), 339–356.

Xiong, H. B., Chen, L. H., & Lin, J. Z. (2006). Smoothed particle hydrodynamics modeling of free surface flow. *Journal of Hydrodynamics, 18*(1), 443–445.

Yu, L., & Batlle, F. (2011). A hybrid method for quasi-three-dimensional slope stability analysis in a municipal solid waste landfill. *Waste Management, 31*(12), 2484–2496.

Zekkos, D., Athanasopoulos, G. A., Bray, J. D., et al. (2010). Large-scale direct shear testing of municipal solid waste. *Waste Management, 30*(8–9), 1544–1555.

Zhan, L. T., Chen, Y. M., & Ling, W. A. (2008). Shear strength characterization of municipal solid waste at the Suzhou landfill, China. *Engineering Geology, 97*(3–4), 97–111.

Chapter 6
SPH Modeling for Flow Behavior of Liquefied Soils

Large ground displacement caused by seismic liquefaction is one of the main reasons for damage to highways, railways, bridges, and other lifeline engineering. Considering the disadvantages of current research methods for large deformation analysis of flowing liquefied soil, this chapter uses the Smoothed Particle Hydrodynamics (SPH) method to carry out numerical simulations for analyzing flow behavior of sandy soil after liquefaction.

6.1 Introduction

6.1.1 Liquefaction Hazards

Among various types of disasters, earthquakes are one of the most serious for human society. There have been many earthquakes worldwide that have been disastrous to lives and property, such as Yugoslavia 1963, Niigata Japan 1964, Alaska USA 1964, Mexico 1985, Armenia Soviet Union 1988, Great Hanshin Japan 1995, and Izmit Turkey 1999 (Zhang et al. 2005). China, located between the Himalayan and Pacific seismic belts, is earthquake-prone and has endured several strong earthquakes. Examples include, Haiyuan 1920 (Ningxia), Xingtai 1966 (Hebei), Haicheng 1975, Tangshan 1976 (the most severe earthquake in the world), Taiwan Chi-chi 1999, and Wenchuan 2008. In recent years, Chinese earthquakes have had strong intensity, high frequency, and shallow epicenters.

Collapse and destruction of structures are major causes of the loss of life and property in an earthquake. These structural phenomena can usually be attributed to insufficient bearing capacity leading to destruction, or foundation failure leading to collapse. Liquefaction induced by the earthquake can greatly reduce the bearing capacity of a foundation. For example, the Xingtai, Haicheng, and Tangshan earthquakes triggered a wide range of liquefaction phenomena, causing serious disaster (Wang et al. 1982; Chen et al. 1999; Yi et al. 2005). During the Ms 8.0 Wenchuan earthquake of 2008 in China, there was widespread liquefaction in Chengdu,

© Springer-Verlag Berlin Heidelberg 2014
Y. Huang et al., *Geo-disaster Modeling and Analysis: An SPH-based Approach*,
Springer Natural Hazards, DOI 10.1007/978-3-662-44211-1_6

Mianyang, and Deyang, and in parts of Leshan, Ya'an, Suining, and other locations (Huang and Jiang 2010). In the Mw 9.0 Tohoku (Japan) earthquake of 2011, liquefaction was widespread in the Tokyo Bay area. As a result, many sections of pavement, manholes, and highway footings were destroyed (Bhattacharya et al. 2011).

Besides reducing the bearing capacity of foundations, liquefied soil may flow as a type of liquid. Flow slides from liquefaction are one of the most serious geo-disasters. During the Tangshan earthquake, the banks of the Douhe, Luanhe, Jiyunhe, Haihe, and Yueyahe rivers slid with sand blasting. This shortened and destroyed the Tangshan Shengli, Yuehe, and Hangu bridges, and more than 10 highway and railway bridges (Chen 2001). After the 1989 Loma Prieta earthquake in San Francisco, several revetment walls moved as much as 20 m because of soil liquefaction (Bartlet and Youd 1995). In the Hanshin earthquake, lateral flow of liquefied foundations damaged port facilities and produced subsidence of caissons toward the sea. The maximum permanent displacement was 5 m, and facilities such as electric, waterways, gas lines, and mobile stations were substantially destroyed (Committee on Earthquake Engineering 1996). Compared with pure liquefaction, flow slides of liquefied soil tend to be more severe and widespread, with catastrophic consequences. In view of this, it is necessary to study large-scale deformation and flow mechanisms of liquefied soils for disaster evaluation and improvement of infrastructure.

6.1.2 Definition and Mechanism of Liquefaction

Liquefaction of sandy soil refers to a pore water pressure increase and effective stress decrease, which transforms soil from solid to liquid under loadings other than static ones. In this process, effective stress (strength) declines and even disappears. The development of liquefaction can be defined as the entire process of soil structural damage, pore water pressure rise, and reduction of strength. Sand blasting, water gushing, flow sliding, buildings floating in soil, and building subsidence are macroscopic phenomena of liquefaction. Thus, the liquefied state is the fully developed result of the liquefaction process and the liquefaction is characteristic of this process under the actions of various factors.

The mechanism of liquefaction of saturated soil can usually be described by the following: (1) Sand boiling—Excess pore pressure is produced by rising groundwater in the soil; when the pore pressure equals or exceeds the weight of upper soil, sand boiling is initiated. (2) Cyclic loading. Cyclic shear loading leads to the contraction and dilatation cycle of the soil volume, such that the corresponding pore water pressure cycle generates intermittent liquefaction and finite deformation. This mechanism mainly pertains to medium-density or relatively dense saturated clay. (3) Flow slide. Under the action of simple shear, the soil volume continues to shrink. The increase of pore water pressure and sudden decline of shear strength causes the soil to flow. This mechanism mainly relates to loose and undrained saturated clay.

The sharp rise of pore water pressure rapidly reduces soil shear strength, possibly to the point of its complete loss. Then, liquefaction triggers large and continuous deformation. If this occurs on flat ground or on a slope of certain inclination, the liquefied soil can induce a wide range of ground flow. This may be defined as regional sand liquefaction.

6.1.3 Current Research on Liquefaction

In accord with the level of development of shear strain, the entire process of liquefaction can be divided into the initial, partially liquefied (① in Fig. 6.1), early liquefied (②), completely liquefied (③), and flow (④) states. Current studies on the large deformation of sand liquefaction focus on the partially liquefied, early liquefied and completely liquefied states, in which the maximum shear strain of the soil reaches only about 20 %. However, for sand liquefaction over a large area, this strain exceeds 100 % in the flow state.

While the shear strain fully develops, the movement of soil material is characterized by distinct fluidization after liquefaction. In such a case, this material can no longer be regarded as a solid, and the nature of deformation is similar to that of a viscous fluid. Research on liquefaction has concentrated on liquefied conditions, and studies of liquefied soil deformation have focused on the initial stages of liquefaction instead of the flow stage. Traditional analysis methods in the framework of solid mechanics have difficulty in treating the soil transition from solid to liquid phase. Therefore, flow processes of liquefied soils as viscous fluids must be analyzed to obtain dynamic behaviors and improve the resistance of infrastructures to seismic events.

Computational Fluid Dynamics (CFD) is often used to simulate flow behavior of soils after liquefaction. For example, Uzuoka et al. (1998) and Hadush et al. (2000)

Fig. 6.1 Development of liquefaction

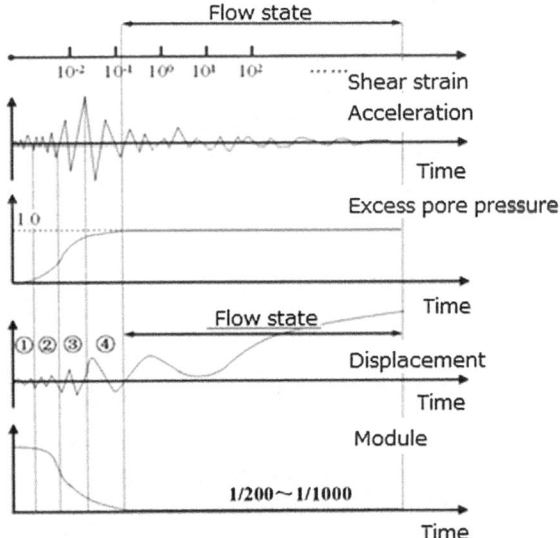

introduced the concept of equivalent viscosity into a Bingham fluid model to solve the problems of unsteady flow and large deformation of liquefied soils. Hadush et al. (2001) and Sawada et al. (2004) used the Cubic Interpolated Pseudo-particle (CIP) method to study large deformation in liquefied soils and reproduce the dynamic behaviors of ground displacement and velocity variations in soil at various depths. Simulations of liquefied soils have mainly been based on traditional CFD methods. These methods determine the free surface according to the ratio of fluid flow through a grid to the volume of a cell, and they track changes in the fluid rather than the movement of particles on the free surface. Although the methods can analyze the deformation and nature of soil flow after liquefaction in certain situations, precise determination of a free surface is often difficult. Since the methods are based on a Euler description in which grids are fixed in model space, it is difficult to deal with deformation boundary conditions and the interface of different phases, especially in an irregularly shaped model. Therefore, a more advanced numerical method is needed to simulate liquefied soil flow processes.

The recently developed technique of Smoothed Particle Hydrodynamics (SPH) has a unique advantage in dealing with the problem of free surfaces, deformation boundaries, and large deformations (Liu and Liu 2003). SPH can readily accommodate large deformation and the flow stage of geo-materials, and is therefore suitable for analysis of liquefied soil flow processes. Liquefied soil can be modeled with two main constituents, pore water and a solid soil skeleton.

Pore water is important in the behavior of liquefied soil. Therefore, for precise analyses, coupling of the pore water and solid skeleton must be considered during SPH simulation of liquefied soils. Pastor et al. (2008) adopted the Zienkiewicz-Biot model to calculate the interaction between water and soil, and proposed an SPH method that may be used in the simulation of a flow-like landslide. Maeda et al. (2006) used a water-soil-air-coupled SPH method to simulate the seepage failure of dykes. Bui et al. (2008) applied the Drucker-Prager model with a non-associated plastic flow rule and the Von-Mises yield criterion to simulate the interaction of soil and structure. Research using coupled SPH simulations of liquefied soils is rare in the literature, but it is essential to conduct the coupling analysis and obtain more precise flow mechanisms of soils after liquefaction.

Previously, we simulated the flowing process of liquefied soil using the SPH method for a Bingham fluid. In that study, the liquefied soil was considered a viscous fluid and interactions between pore water and the soil skeleton were not considered. Based on prior research, this section introduces the mixture theory of Biot (1941) into the SPH method to establish a soil–water-coupled SPH model for the flowing analysis of liquefied soil.

6.1.4 Soil–Water-Coupling Algorithm

Liquefied soil is composed of water and soil particles. The coupling effect between the liquid and solid phases may be significant in the flow behavior of liquefied soil. The liquid phase (water) is considered an incompressible fluid. In the traditional SPH model for a fluid, the total stress tensor is typically divided into two parts, i.e., an isotropic hydrostatic pressure p and deviatoric shear stress τ:

$$\sigma^{\alpha\beta} = -p\delta^{\alpha\beta} + \tau^{\alpha\beta}, \tag{6.1}$$

where $\delta^{\alpha\beta}$ is Kronecker's delta, $\delta^{\alpha\beta} = 1$ if $\alpha = \beta$, and $\delta^{\alpha\beta} = 0$ while $\alpha \neq \beta$.

In the SPH model, the liquid pressure can be calculated by the equation of state for quasi-incompressible fluids, as proposed by Batchelor (1967):

$$p_d = p_0 \left(\left(\frac{\rho^f}{\rho_0^f} \right)^\gamma - 1 \right), \tag{6.2}$$

where ρ^f is the density of liquid calculated by the continuity equation and ρ_0^f is the reference density. The smaller the γ, the greater the compressibility. Following Monaghan (1994), γ was set to 7.0 for effective simulation of an incompressible fluid.

An elastic model was selected as the constitutive model of the solid phase. Therefore, the relationship between stress rate and strain rate can be described as:

$$\frac{d\sigma_{ij}}{dt} = D_{ijkl}^e \cdot \varepsilon_{kl}^e. \tag{6.3}$$

For plane strain problems,

$$D_{ijkl}^e = \begin{bmatrix} \lambda + 2\mu & \lambda & \lambda & 0 \\ \lambda & \lambda + 2\mu & \lambda & 0 \\ \lambda & \lambda & \lambda + 2\mu & 0 \\ 0 & 0 & 0 & \mu \end{bmatrix}, \tag{6.4}$$

$$\text{where} \quad \lambda = \frac{\nu E}{(1+\nu)(1-2\nu)}, \quad \mu = \frac{E}{2(1+\nu)}. \tag{6.5}$$

Here, E is Young's modulus and ν is the Poisson ratio.

The frictional force is used as the force of interaction in the SPH model, considering the effects of porosity n, permeability k, and the velocity difference between the two phases. According to the mixture theory of Biot (1941), this force can be determined from the following equations:

$$\mathbf{v}^{sf} = \sum_{j=1}^N m_j \frac{\mathbf{v}^f}{\rho_j} W_{ij}, \quad \mathbf{R}^{sf} = n^2 \frac{\rho_f g}{k} (\mathbf{v}^{sf} - \mathbf{v}^s), \tag{6.6}$$

$$\mathbf{v}^{fs} = \sum_{j=1}^N m_j \frac{\mathbf{v}^s}{\rho_j} W_{ij}, \quad \mathbf{R}^{fs} = n^2 \frac{\rho_f g}{k} (\mathbf{v}^{fs} - \mathbf{v}^f). \tag{6.7}$$

Here, $\mathbf{v}^{sf}$ and $\mathbf{v}^{fs}$ are the summarized velocities of the fluid and solid phases, respectively; $\mathbf{R}^{sf}$ is the interaction force exerted on the solid phase by the fluid phase; $\mathbf{R}^{fs}$ is the interaction force exerted on the fluid phase by the solid phase; n is porosity; ρ_f is the density of the fluid phase; g is gravitational acceleration; and k is the permeability coefficient.

6.2 Physical Model Test of Liquefied Soil Flow

Based on an existing model testing installation, we designed and conducted a physical flow model test of liquefied sand to obtain the dynamic behaviors. Then, the model test was simulated using SPH to compare the numerical and test results.

6.2.1 Model Test Device

The experimental device (Fig. 6.2) designed by the authors has five parts: a motor with a reduction gear, a slider crank, a shaking table, a model box, and a base. The motor and reduction gear produce rotation at a rate of 140 revolutions per minute. The slider crank converts the rotation into a back-and-forth movement, which in turn produces horizontal vibration of frequency 1–10 Hz. The model box containing saturated sand has dimensions of 98 × 35 × 34 cm and is placed on the vibrating base. The model box is made of transparent organic glass, with grids etched on it to facilitate observation of flow configurations in the liquefied sand. Pressure and velocity can be measured by probes installed in the model box and a high-speed camera system.

The initial condition of the model under testing is shown in Fig. 6.3. First, the left side of the model box is separated from the right side by a baffle, without leakage. Then, water is injected into the left side of the model box, to a height about one-third that of the box. After that, sandy soil is spread by the hopper and completely saturated. The soil sample has an area of 32 × 15 cm. Vibrations of 1 Hz are imposed on the shaking table to liquefy the sand sample. When the excess pore pressure maximizes (Fig. 6.4), the baffle is removed and the liquefied sand flows under gravity. The high-speed camera record showed that the flow lasted 6 s.

Fig. 6.2 Physical model test device (reprinted from Huang et al. (2011) with permission from Springer)

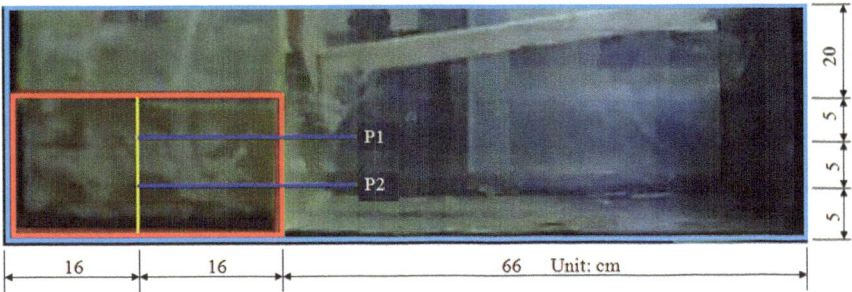

Fig. 6.3 Initial conditions of physical model test (reprinted from Huang et al. (2011) with permission from Springer)

Fig. 6.4 Excessive pore pressure during model test (reprinted from Huang et al. (2011) with permission from Springer)

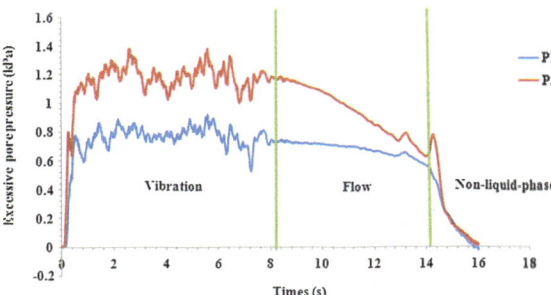

6.2.2 SPH Simulation of Model Test

Accompanying the physical model test, a corresponding numerical model was established. In this model, there were 2,904 total particles, 1920 for the soil and 984 for the boundary, with particle spacing of 1 cm. As in the model test, the soil sample could be deformed both horizontally and vertically, with gravitation applied only in the vertical direction. Parameters used in the SPH simulation of the physical model are listed in Table 6.1. Density and triaxial compression tests were conducted to obtain the density, frictional angle, and cohesion of the soil sample. The equivalent viscosity coefficient can be determined by theoretical formulae, according to the frictional angle, cohesion, and shear-strain rate in the model test. It took almost 2 hrs to complete the numerical simulation. The deformations obtained through the physical model and SPH simulation are shown in Fig. 6.5. Figure 6.6 shows the velocity distribution of liquefied sand. This figure shows that velocity vectors of many particles near the ground surface are directed upward. The main cause for this is downward

Table 6.1 Parameters in SPH simulation of physical model test (reprinted from Huang et al. (2011) with permission from Springer)

Density	$\rho(\text{kg/m}^3)$	1600
Equivalent viscosity coefficient	η (Pa·s)	1.0
Total steps	n	6000
Unit time step	Δt (s)	0.001

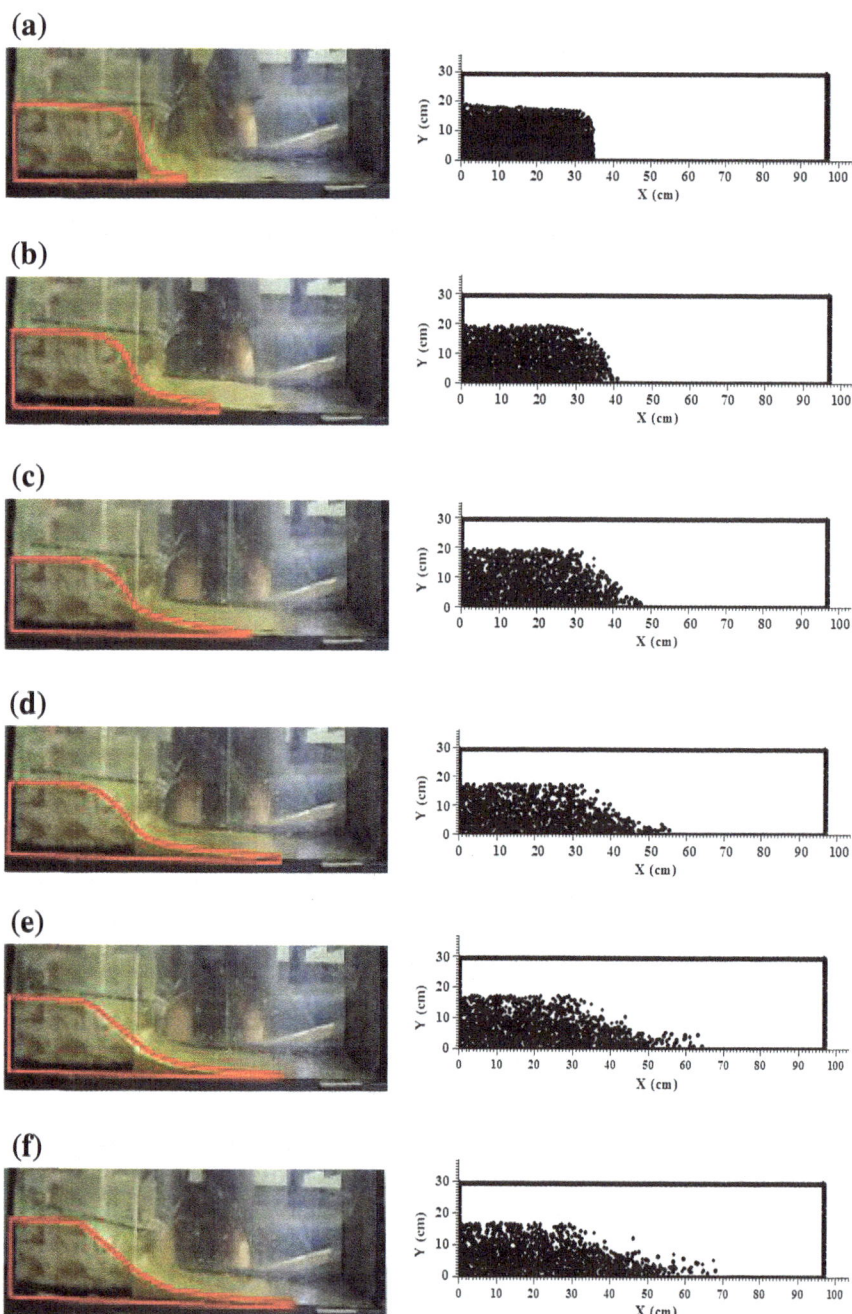

Fig. 6.5 Configuration of model test and SPH simulation (reprinted from Huang et al. (2011) with permission from Springer). **a** $t = 1$ s, **b** $t = 2$ s, **c** $t = 3$ s, **d** $t = 4$ s, **e** $t = 5$ s, **f** $t = 6$ s

Fig. 6.6 Velocity
distributions within liquefied
sand (reprinted from Huang
et al. (2011) with permission
from Springer). **a** $t = 1$ s, **b**
$t = 2$ s, **c** $t = 3$ s, **d** $t = 4$ s, **e**
$t = 5$ s, **f** $t = 6$ s

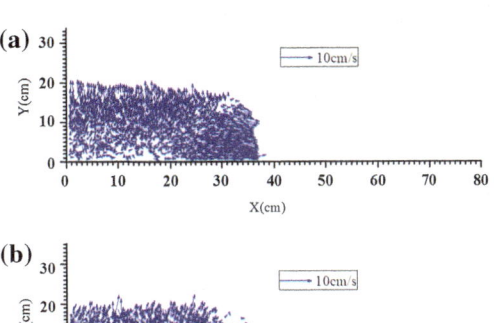

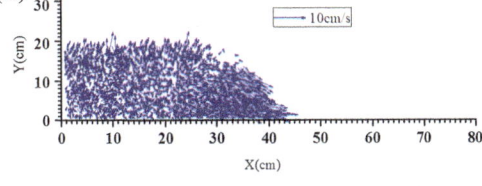

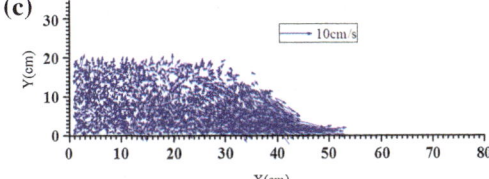

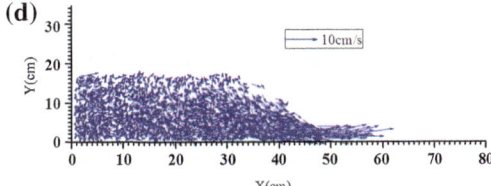

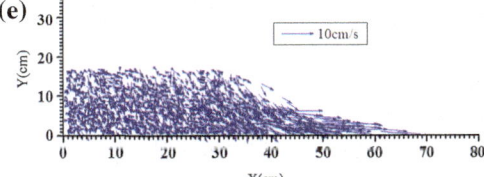

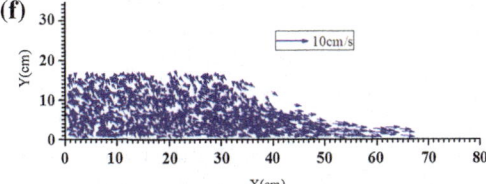

particle flowed at the beginning; boundary particles produced an interaction force in the opposite direction to prevent soil particles from penetrating.

The SPH results indicate that the flow was significant for the first 3 s after baffle removal. After that, the liquefied sand at the free surface boundary continued to flow. However, because of the dissipation of excess pore pressure, shear strength increased and the shear-strain rate decreased. The flow ceased at 6 s after baffle removal.

There were some deviations between the SPH simulation and physical model test, particularly for the velocity and sliding distance on the right side of the free surface. The main reason for this is that the parameters in the SPH simulation were derived from the triaxial test of the soil sample. The results of that test were influenced by many factors, such as the test conditions, interaction of water and air, strain rate, and others. These factors differ from the actual flow test and are not considered in the numerical method. However, the numerical results largely correspond with those parameters.

6.2.3 Coupled SPH Simulation of the Model Test

The model test described above was simulated by the coupled model. The numerical model had 528 water, 528 soil, and 564 boundary particles. Water and soil particles could occupy the same positions, and flowed under gravity. Simulation parameters are given in Table 6.2. These were obtained from the laboratory test results for the soil sample. During the flow of liquefied soil, the viscosity of liquefied soil and the shear module both decrease with increasing shear strain (Hadush et al. 2000). Therefore, the elastic module decayed in the SPH simulation. It took 7.2 h to complete this simulation on a personal desktop computer with Core i5 760, 4 GB RAM, and Windows 7. The simulated configurations of flowing liquefied sand are shown in Fig. 6.7 and compared with the test configurations. The simulated results have good agreement with the model test. Figures 6.8 and 6.9 show the SPH-simulated velocity vectors for soil and water, respectively.

From the velocity vectors, we see that the flow velocity was significant for the first 3 s after baffle removal. After that, the liquefied sand at the free surface boundary continued to flow, but because of the dissipation of excess pore pressure, shear strength increased and the shear-strain rate decreased. The flow ceased 6 s after baffle removal. The overall flow process as determined by the SPH analysis is

Table 6.2 Parameters in SPH simulation of physical model test using coupled model (reprinted from Huang et al. (2013) with permission from Springer)

Density of water $\rho(kg/m^3)$	Density of soil particles $\rho(kg/m^3)$	Porosity e	Particle spacing $L(m)$
1000	2681	0.60	0.01
Young's modulus E (MPa)	Poisson ratio ν	Unit time step $\Delta t(s)$	Total steps n
10	0.3	0.00003	200000

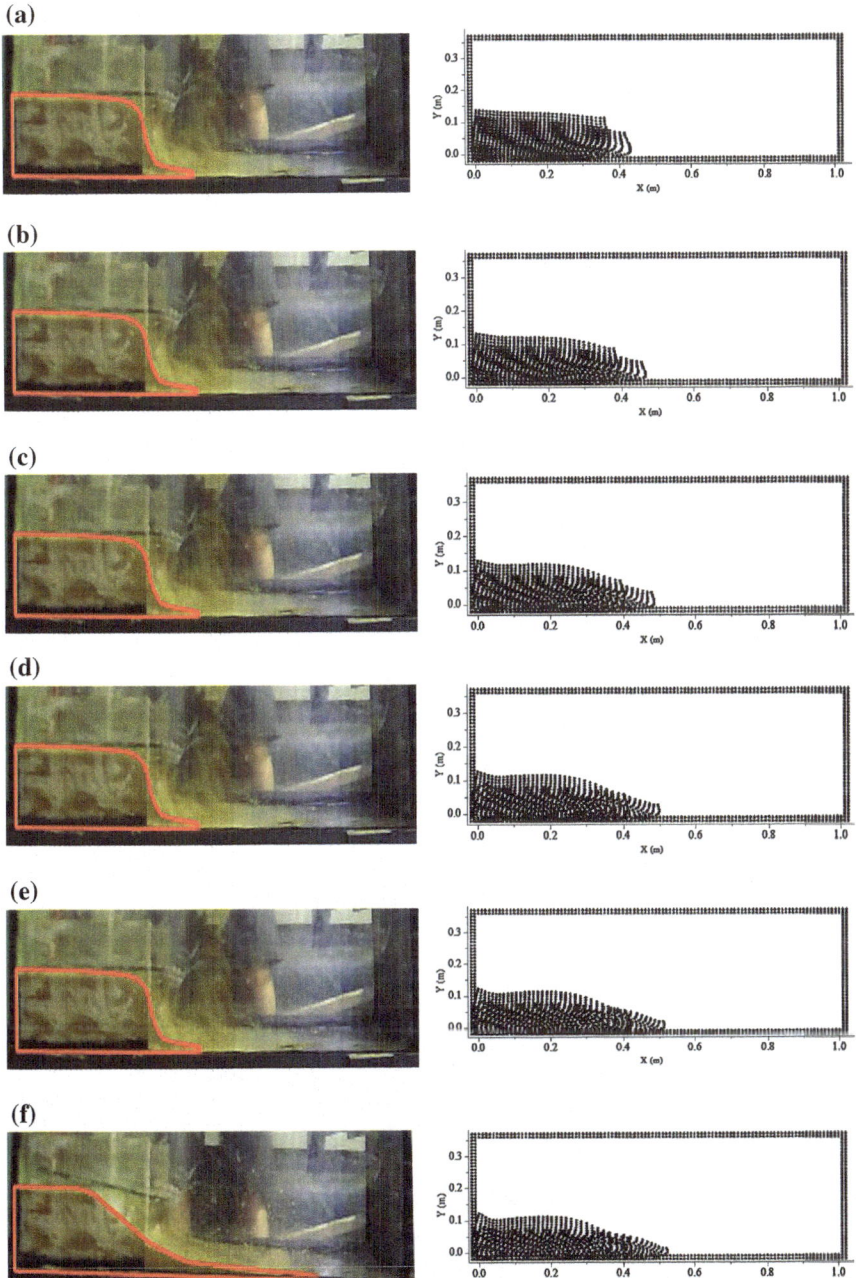

Fig. 6.7 Model test and SPH-simulated configurations (reprinted from Huang et al. (2013) with permission from Springer). **a** $t = 1$ s, **b** $t = 2$ s, **c** $t = 3$ s, **d** $t = 4$ s, **e** $t = 5$ s, **f** $t = 6$ s

Fig. 6.8 SPH-simulated velocity vectors for soil (reprinted from Huang et al. (2013) with permission from Springer). **a** $t = 1$ s, **b** $t = 2$ s, **c** $t = 3$ s, **d** $t = 4$ s, **e** $t = 5$ s, **f** $t = 6$ s

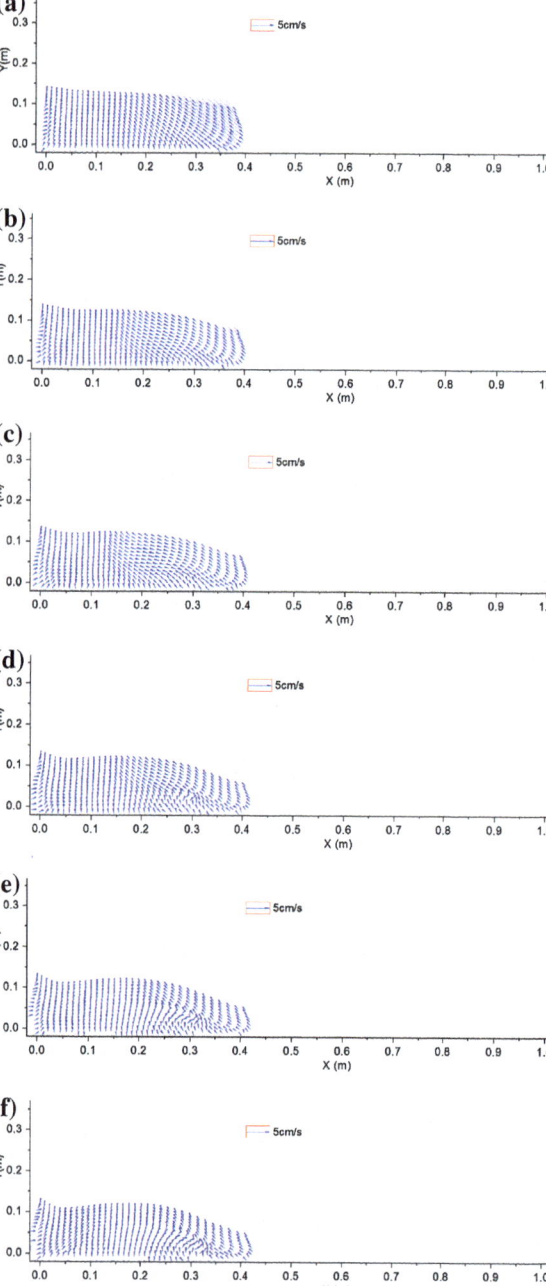

Fig. 6.9 SPH-simulated velocity vectors for water (reprinted from Huang et al. (2013) with permission from Springer). **a** $t = 1$ s, **b** $t = 2$ s, **c** $t = 3$ s, **d** $t = 4$ s, **e** $t = 5$ s, **f** $t = 6$ s

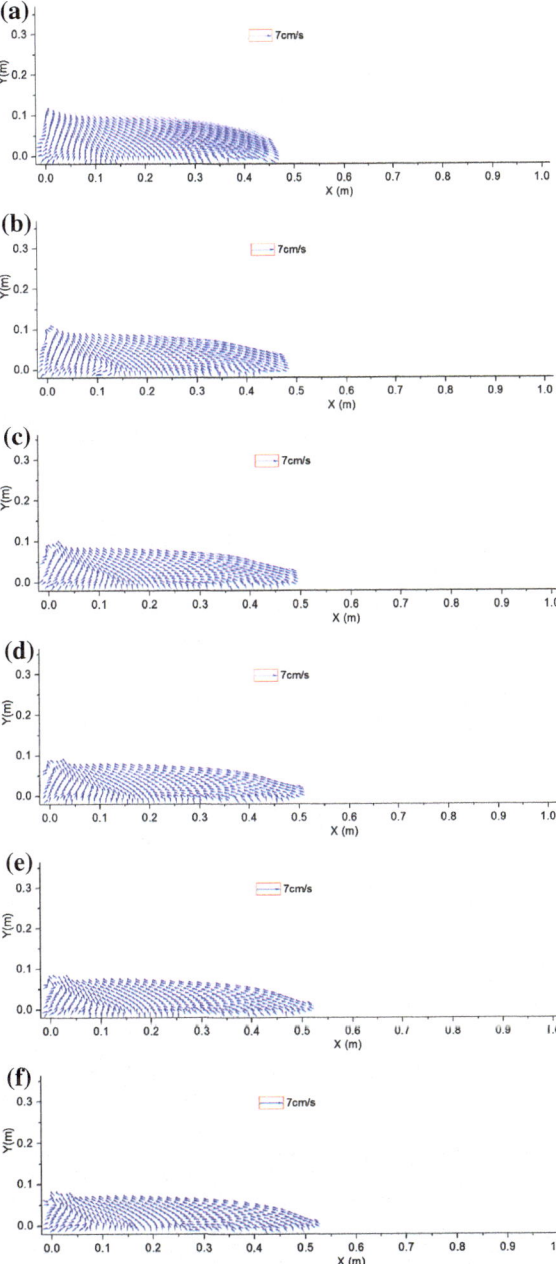

characterized by water velocities that are greater than those of the soil. As a result, the flow distance of the water was greater than that of the soil, and water seeped out of the soil. In the model test, water also seeped from the soil and flowed out in front at the late flow stage. This phenomenon could explain the reduced velocity vectors; i.e., water seeped from the soil and then excessive pore pressure dissipated.

6.3 Shaking Table Test of Liquefied Soil Flow

6.3.1 Shaking Table Test Device and Results

A shaking table test designed by Hamada et al. was simulated using the SPH model. The physical test device is shown in Fig. 6.10. For a detailed description of the device, please refer to the original paper (Hamada et al. 1994). In this test, the soil sample became completely liquefied after 9 s of cyclic loading. After suspending the vibration, the model box was tilted at an incline of 4.2 % (2.4°). The liquefied soil then started to flow from the left side to right side in the model box, which continued for ~11 s. Flow velocities and horizontal displacements were measured by probes at three different depths in the soil sample, annotated as 1, 2, and 3.

As shown in Fig. 6.11, horizontal displacements at the three depths began to increase at 9 s. They reached maxima by 11 s and remained constant thereafter. Figure 6.12 shows velocity time series at the three depth points, from which it is clear that the evolution of velocity had two stages: an early increase toward a peak and a subsequent decrease.

6.3.2 SPH Simulation of Shaking Table Test

The size of the SPH numerical model was the same as the physical model box, and particle spacing was 2 cm. The numerical model had a total of 2,760 particles,

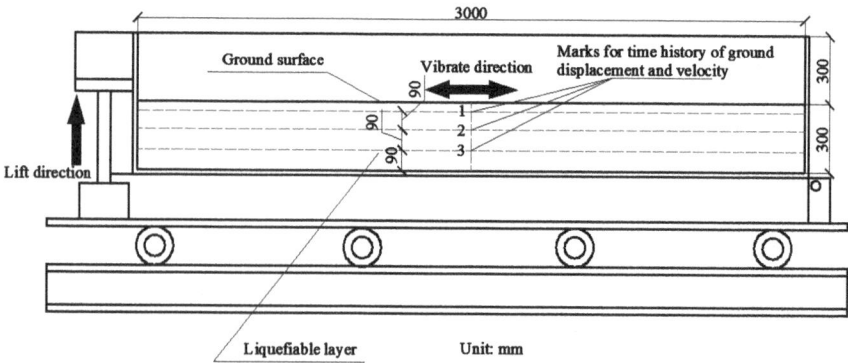

Fig. 6.10 Shaking table test device (reprinted from Huang et al. (2011) with permission from Springer)

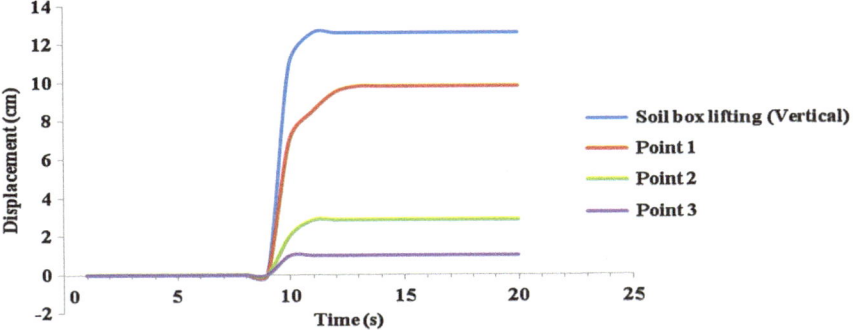

Fig. 6.11 Time series of horizontal displacements at measuring points (reprinted from Huang et al. (2011) with permission from Springer)

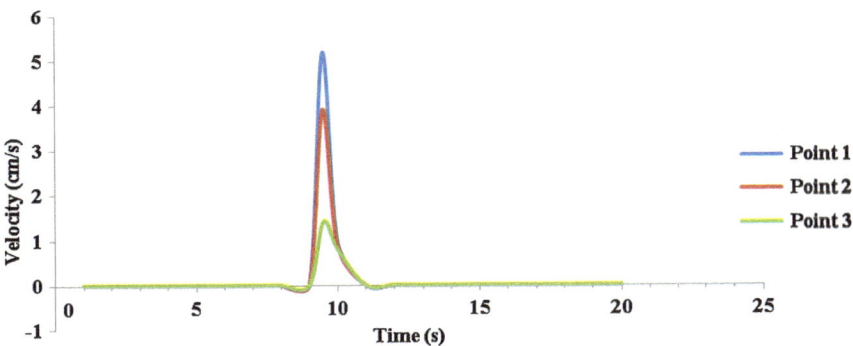

Fig. 6.12 Time series of velocities at measuring points (reprinted from Huang et al. (2011) with permission from Springer)

Table 6.3 Parameters in SPH simulation of shaking table test (reprinted from Huang et al. (2011) with permission from Springer)	Density	$\rho(\mathrm{kg/m^3})$	1800
	Equivalent viscosity coefficient	$\eta'(\mathrm{Pa \cdot s})$	2100
	Total steps		5500
	Unit time step	s	0.002

2,336 for the soil sample and another 424 for the boundary. Under the action of gravity, the soil sample could deform both horizontally and vertically. Parameters for the SPH simulation of the shaking table test are listed in Table 6.3. The soil sample in the model test was sand from the Enshunada coast in Japan, with relative density of 23 %. Hwang et al. (2006) studied the relationship between viscosity and relative density. In this SPH simulation, the viscosity was selected as 2,100 Pa·s. This viscosity is much greater than that used in previous physical model tests because the shear-strain rate in our experiment was much smaller, and

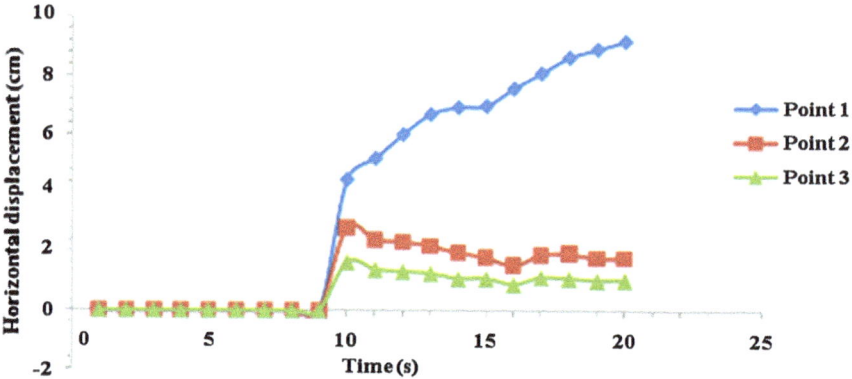

Fig. 6.13 SPH-simulated time series of horizontal displacements (reprinted from Huang et al. (2011) with permission from Springer)

the viscosity sharply increases with decreasing shear-strain rate (Hadush et al. 2000; Hwang et al. 2006). It took almost 1.5 hrs to run the code and complete the SPH simulation. Time histories of horizontal and vertical displacements and velocities of three particles, which were at the same locations as the three aforementioned measuring points, were output to analyze flow mechanisms of the liquefied sand (Figs. 6.13, 6.14, and 6.15).

Figure 6.13 shows that the horizontal displacement of point 1 increased rapidly from 9 to 19 s and then reached a stable state. However, the horizontal displacements of points 2 and 3 increased until 10 s, decreased slightly at 11 s, and then attained a stable state. These displacement variations correspond with those observed in the shaking table test (Fig. 6.11). In addition, the horizontal displacements at all three points during the stable stage (Table 6.4) correlate well with those measured in the test.

Figure 6.14 shows the time histories of vertical displacements. The displacements at the three measuring points increased at 9 s and reached a stable state 2 s later. In the stable state, the vertical displacements were about 1.1, 0.9, and 0.6 cm, respectively. Subsidence at the upstream end of the model box is also shown in the figure, with a maximum of 3.7 cm. Hamada et al. (1994) proposed empirical formulae (Eqs. 6.1 and 6.2) to estimate subsidence at the upstream end of a soil box:

$$x = L - h\sqrt{\gamma_c/\theta}, \tag{6.7}$$

$$\delta = \theta \cdot (L - x), \tag{6.8}$$

where δ is the subsidence, h is soil thickness, L is the half length of the soil box, θ is the gradient, and γ_c is the critical shear strain. According to the equations and

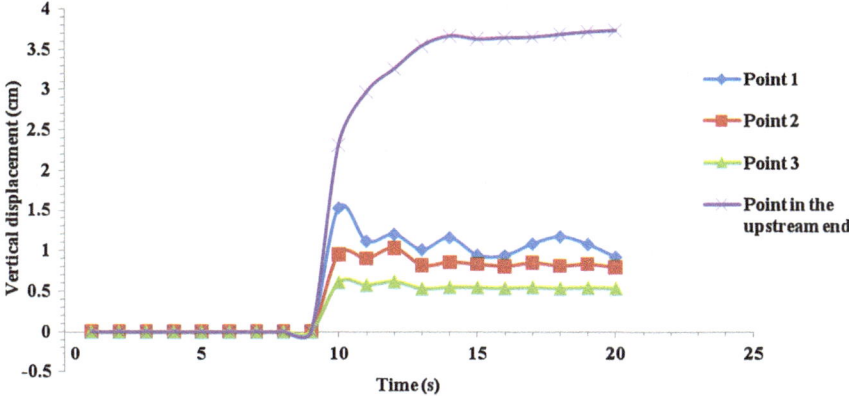

Fig. 6.14 SPH-simulated time series of vertical displacements (reprinted from Huang et al. (2011) with permission from Springer)

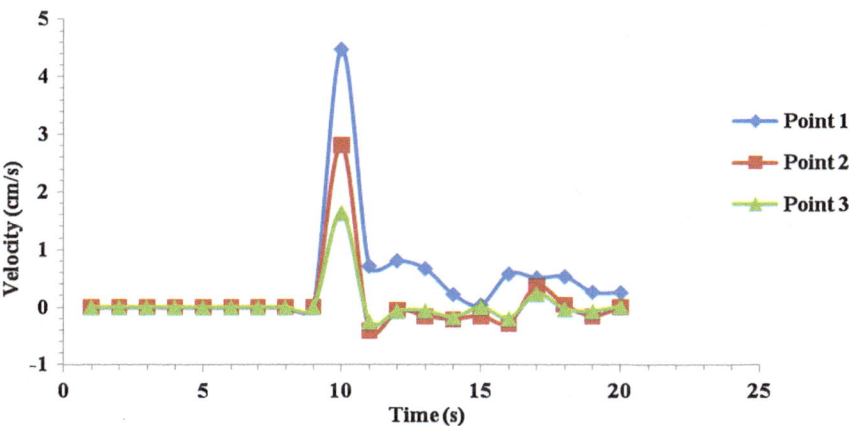

Fig. 6.15 SPH-simulated time series of velocities (reprinted from Huang et al. (2011) with permission from Springer)

Table 6.4 Comparison of horizontal displacements in the stable state between SPH and shaking table test (reprinted from Huang et al. (2011) with permission from Springer)

Measuring points	SPH simulation (cm)	Shaking table test (cm)
1	9.8	9.6
2	2.4	2.5
3	1.0	1.0

Table 6.5 Comparison of maximum velocities between SPH and shaking table test (reprinted from Huang et al. (2011) with permission from Springer)

Measuring points	SPH simulation (cm/s)	Shaking table test (cm/s)
1	4.6	5.2
2	2.9	3.7
3	1.5	1.3

related data, the vertical displacement was calculated at 3.98 cm, which is very close to the SPH-simulated result of 3.7 cm. The SPH-simulated subsidence was also very near the calculated value.

From the time histories of velocities shown in Fig. 6.15, points 1, 2, and 3 all approached maximum velocities around 10 s. Magnitudes of the velocities (Table 6.5) are slightly smaller than the test results. The velocity time series was similar to the physical test curves, and could also be divided into two stages: the early increase toward a peak followed by a decrease. In the SPH simulation, there was slight oscillation in the motion of particles. This produced deviations between the simulated and test results. Even so, the variations of displacement and velocity with time in the numerical simulations largely agreed with physical test results. This indicates that SPH simulations could satisfactorily reproduce flow processes of liquefied soils and constrain temporal variations of horizontal and vertical displacement and velocity.

6.4 Flow Behavior Analysis of a Liquefied Embankment

6.4.1 Situation of Embankment Failure

To extend the application of the coupled SPH model to an actual engineering case of liquefied soil flow, a typical liquefaction-induced flow slide was simulated. On October 4, 1983, an embankment of Road No. 352 in northern Sweden (Fig. 6.16) failed from liquefaction (Ekstrom and Olofsson 1985). In this case, an impoundment reservoir near the embankment caused the water table to rise,

Fig. 6.16 Embankment before and after liquefaction failure (reprinted from Huang et al. (2013) with permission from Springer)

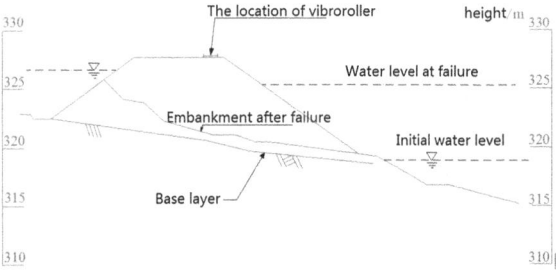

saturating the soil of the embankment. Cracks then appeared on the embankment surface. Subsequently, more cracks developed on the road as a result of repair operations involving a tractor and roller working there. These cracks facilitated the infiltration of water into the embankment. Vibrations of the tractor and roller caused liquefaction, and the liquefied soils then failed and flowed into the bottom of the reservoir.

According to a site investigation by Ekstrom and Olofsson (1985), the distance from the embankment to a riverbed was about 60 m. The embankment was composed of well-graded sand built up via a wet construction process. Because of the glacial source of the sediment, the sand basement was composed of angular quartz and feldspar grains. Average particle size (D_{50}) of the sand was 0.11 m. The uniformity coefficient (Uc) was 17.0, with soil particle density of 2.75 g/cm^3.

6.4.2 Coupled SPH Simulation of Embankment Failure

A numerical model of the embankment was established for SPH analysis. In this model, the embankment was regarded as movable saturated soil, and the base layer was assumed a discrete boundary without flow. To simplify the problem, edges of the embankment in the model were segmented in straight lines. There were 1,576 particles in total. Both water and soil particle numbers were 458. The rest were boundary particles that did not move during simulation, while water and soil flowed under the action of gravity. The parameters in Table 6.6 were derived from field survey data of Ekstrom and Olofsson (1985).

Figure 6.17 shows the simulated configurations at typical time steps. The flow of liquefied soil was fast in the initial stage and then slowed to a stop in the last stage. A comparison between the SPH simulation and aforementioned field survey data is shown in Fig. 6.18. The simulated flow distance and configuration correspond closely with the site investigation. This supports the use of the SPH model to simulate both the flow process and runout distance in real flow events induced by soil liquefaction.

Table 6.6 Parameters used in SPH analysis of embankment failure in Sweden (reprinted from Huang et al. (2013) with permission from Springer)

Density of water ρ (kg/m^3)	Density of soil particles ρ (kg/m^3)	Porosity	Particle spacing L (m)
1000	2750	0.41	0.5
Young's modulus (*initial*) E (MPa)	Poisson ratio ν	Unit time step Δt	Total steps n
7.5	0.3	0.0002 s	600000

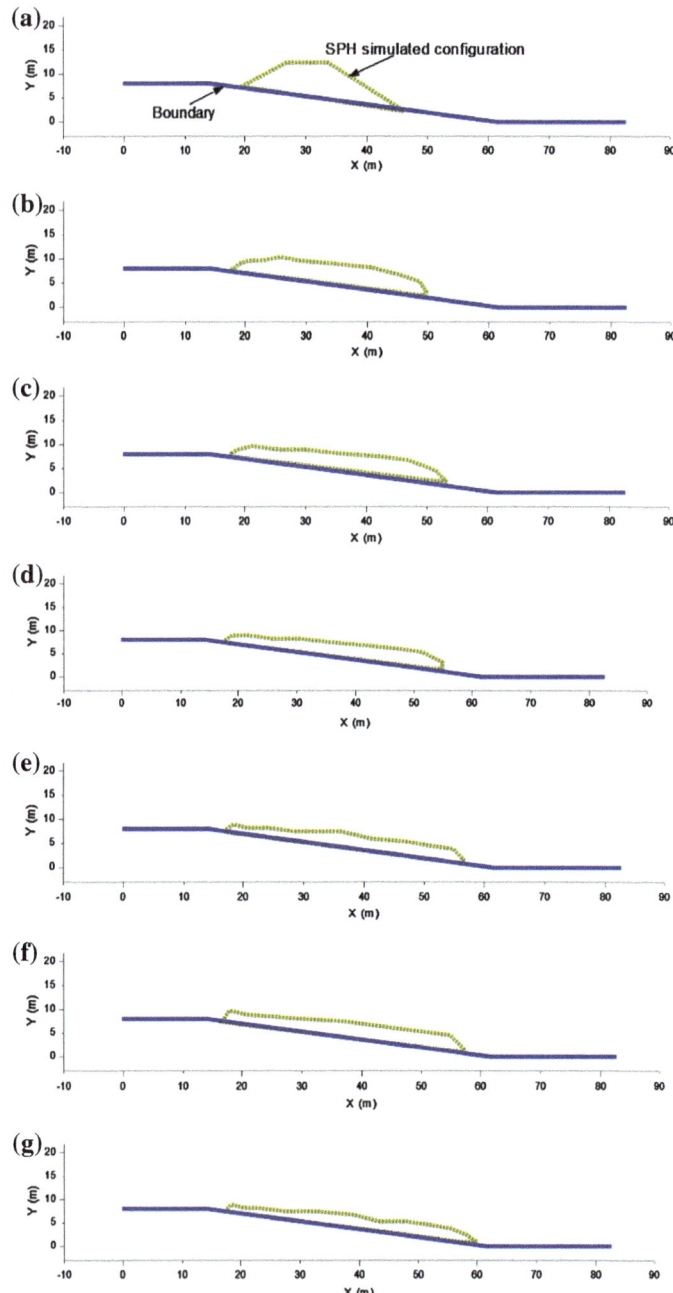

Fig. 6.17 SPH-simulated configurations at 20-s intervals (reprinted from Huang et al. (2013) with permission from Springer). **a** $t = 0$ s, **b** $t = 20$ s, **c** $t = 40$ s, **d** $t = 60$ s, **e** $t = 80$ s, **f** $t = 100$ s, **g** $t = 120$ s

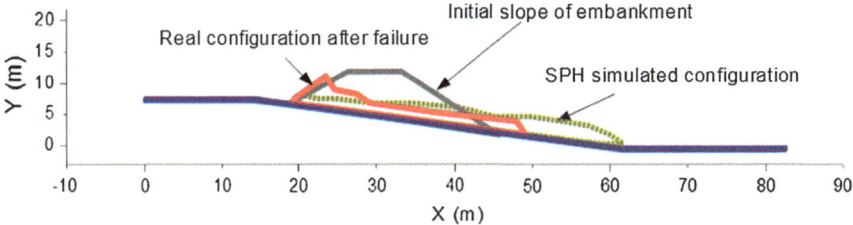

Fig. 6.18 Comparison between SPH simulation and field survey data (reprinted from Huang et al. (2013) with permission from Springer)

6.5 Summary

(1) Liquefaction hazards across the world were summarized and the implications of related research were given from a standpoint of earthquake proofing. Existing research was reviewed, from which the disadvantages of current analysis methods were presented.

(2) SPH models with and without soil–water-coupling algorithm were used to simulate a physical model test of liquefied soil flow. For both cases, simulated configurations, velocities, and flow distances were compared with the model test.

(3) Based on shaking table tests by other scholars, SPH simulations were conducted to determine dynamic behaviors (e.g., flow configuration, velocity, and distance) of liquefied soils. These behaviors can be used to validate the application of the SPH model.

(4) The soil–water-coupled SPH model was successfully used to simulate an embankment failure in northern Sweden, via comparison to a site investigation. The estimated horizontal displacements and final configurations agreed well with survey data.

The SPH models have been shown to satisfactorily reproduce the flow processes of liquefied soils and to estimate the horizontal displacement and velocity of soils after liquefaction. The observed dynamic behavior can benefit structural design in seismic liquefaction zones. As a result, seismic safety attributes of structures can be improved.

References

Batchelor, G.K. (1967). An introduction to fluid dynamics. UK: Cambridge University Press.

Bartlet, S. F., & Youd, T. L. (1995). Empirical prediction of liquefaction-induced lateral spread. *Journal of Geotechnical Engineering, ASCE, 121*(4), 316–329.

Bhattacharya, S., Hyodo, M., Goda, K., Tazoh, T., & Taylor, C. A. (2011). Liquefaction of soil in the Tokyo Bay area from the 2011 Tohoku (Japan) earthquake. *Soil Dynamics and Earthquake Engineering, 31*(11), 1618–1628.

Biot, M. A. (1941). General theory of three-dimensional consolidation. *Journal of Applied Physics, 12*, 152–164.

Bui, H. H., Fukagawa, R., Sako, K., & Ohno, S. (2008). Lagrangian meshfree particles method (SPH) for large deformation and failure flows of geomaterial using elastic-plastic soil constitutive model. *International Journal for Numerical and Analytical Methods in Geomechanics, 32*(12), 1527–1570.

Chen, W. H. (2001). Slipping disaster induced by seismic liquefaction. *Journal of Natural Disasters, 10*(4), 88–93. (in Chinese).

Chen, W. H., Sun, J. P., & Xu, B. (1999). Survey of seismic research on thermal power plant. *World Information on Earthquake Engineering, 15*(1), 16–24. (in Chinese).

Committee on Earthquake Engineering. (1996). *The 1995 Hyogoken-Nanbu earthquake: investigation into damage to civil engineering structures.* Tokyo: Japan Society of Civil Engineers.

Ekstrom, A., & Olofsson, T. (1985). Water and frost-stability risks for embankments of fine-grained soils. *Proceedings of Symposium on Failures in Earthworks* (pp. 155–166). London, UK: Institution of Civil Engineers, London, Telford (Thomas) Ltd.

Hadush, S., Yashima, A., & Uzuoka, R. (2000). Importance of viscous fluid characteristics in liquefaction induced lateral spreading analysis. *Computers and Geotechnics, 27*(3), 199–224.

Hadush, S., Yashima, A., Uzuoka, R., et al. (2001). Liquefaction induced lateral spread analysis using the CIP method. *Computers and Geotechnics, 28*(8), 549–574.

Hamada, M., Sato, H., & Kawakami, T. (1994). A consideration of the mechanism for liquefaction-related large ground displacement In *Proceedings from the Fifth U.S.–Japan Workshop on Earthquake Resistant Design of Lifeline Facilities and Countermeasures Against Soil Liquefaction, Salt Lake City, USA* (pp. 217–232).

Huang, Y., & Jiang, X. M. (2010). Field-observed phenomena of seismic liquefaction and subsidence during the 2008 Wenchuan earthquake in China. *Natural Hazards, 54*(3), 839–850.

Huang, Y., Zhang, W. J., Dai, Z. L., & Xu, Q. (2013). Numerical simulation of flow processes in liquefied soils using a soil-water-coupled smoothed particle hydrodynamics method. *Natural Hazards, 69*(1), 809–827.

Huang, Y., Zhang, W.J., Mao, W.W., & Jin, C. (2011). Flow analysis of liquefied soils based on smoothed particle hydrodynamics. *Natural Hazards, 59*(3), 1547–1560.

Hwang, J. I., Kim, C. Y., Chung, C. K., & Kim, M. M. (2006). Viscous fluid characteristics of liquefied soils and behavior of piles subjected to flow of liquefied soils. *Soil Dynamics and Earthquake Engineering, 26*(2–4), 313–323.

Liu, G. R., & Liu, M. B. (2003). *Smoothed particle hydrodynamics: a mesh-free particle method.* Singapore: World Scientific Press.

Maeda, K., Sakai, H., & Sakai, M. (2006). Development of seepage failure analysis method of ground with smoothed particle hydrodynamics. *Journal of Structural and Earthquake Engineering, JSCE, Division A, 23*(2), 307–319.

Monaghan, J. J. (1994). Simulating free surface flows with SPH. *Journal of Computational Physics, 110*(2), 399–406.

Pastor, M., Haddad, B., Sorbino, G., Cuomo, S., & Drempetic, V. (2008). A depth-integrated, coupled SPH model for flow-like landslides and related phenomena. *International Journal for Numerical and Analytical Methods in Geomechanics, 33*(2), 143–172.

Sawada, K., Moriguchi, S., Yashima, A., Zhang, F., & Uzuoka, R. (2004). Large deformation analysis in geomechanics using CIP method. *JSME International Journal, 47*(4), 735–743.

Uzuoka, R., Yashima, A., Kawakami, T., & Konrad, J. M. (1998). Fluid dynamics based prediction of liquefaction induced lateral spreading. *Computers and Geotechnics, 22*(3–4), 243–282.

Wang, K. L., Sheng, X. B., Cai, L. Y., et al. (1982). Characteristic of liquation of soil in the areas with various intensities during Tangshan earthquake and criteria for recognition of liquation. *Seismology and Geology, 4*(2), 59–70. (in Chinese).

Yi, R. Y., Liu, Y. M., Li, Y. L., et al. (2005). The relation between earthquake liquefaction and landforms in Tangshan region. *Research of Soil and Water Conservation, 12*(4), 110–112. (in Chinese).

Zhang, G. M., Zhang, X. D., Wu, R. H., et al. (2005). Retrospect of earthquake forecast and prospect. *Recent Developments in World Seismology, 5*, 39–53. (in Chinese).

Chapter 7
SPH Modeling for Propagation of Flow-like Landslides

A deadly earthquake measured at 8.0 Ms occurred at 14:28 on 12 May 2008 in Wenchuan County, Sichuan Province, China, destroying many buildings and causing numerous casualties. Many secondary geologic disasters were triggered by this earthquake, e.g., debris flows, flow-like landslides, debris, and rock avalanches. These disasters also caused great damage and many casualties.

In this chapter, fluidization characteristics of flow-like landslides were examined through Smoothed Particle Hydrodynamics (SPH) simulations of runout for such landslides. Results are as follows.

1. Information on the epicenter, hypocenter, and tectonics of affected areas and disasters of the Wenchuan earthquake is summarized from investigations in affected areas and a search of related literature.
2. A systematic summary and analysis of existing research methods for landslides induced by strong earthquakes are provided, from which problems and disadvantages of traditional numerical methods are presented. On this basis, advantages of the SPH method for the runout analysis of flow-like landslides triggered by strong earthquakes are described.
3. The 2D model established in previous chapters is developed into a 3D version, and a no-slip boundary condition is incorporated to model the effect of a solid boundary on slope movement.
4. For Donghekou landslide in Qingchuan County, shear-strength parameters of a soil sample are obtained through on-site investigation and related experiments, which provide data support to reveal the flow mechanism of a flow-like landslide.
5. SPH simulations of three typical flow-like landslides, Donghekou, Wangjiayan, and Tangjiashan, are conducted, and the application of the SPH method is validated.

The work described in this chapter investigated the flow of earthquake-induced landslides using the SPH model. This numerical modeling captured the fundamental dynamic behavior of these flow-like landslides and gives useful results for hazard assessment and site selection toward reconstruction in earthquake-prone areas.

© Springer-Verlag Berlin Heidelberg 2014   155
Y. Huang et al., *Geo-disaster Modeling and Analysis: An SPH-based Approach*,
Springer Natural Hazards, DOI 10.1007/978-3-662-44211-1_7

7.1 Introduction

7.1.1 Earthquake-triggered Flow-Like Landslides

The aforementioned Wenchuan earthquake had a strong intensity, shallow hypo-center depth, and tremendous destructive power. It affected most regions in China, save for Jilin, Heilongjiang, and Xinjiang provinces. Its epicenter was in Yingxiu town of Wenchuan County. There was nearly 10 km² of serious earthquake effects. As of 8 October 2008, 69,229 people had perished from the earthquake, with 17,923 missing and 374,643 injured. The direct economic loss was 845.1 billion Yuan.

The Wenchuan earthquake triggered a large number of secondary geologic disasters. Regarding damage of the Wenchuan earthquake and induced geologic hazards, Huang and Li (2009) found the following. The earthquake had a large magnitude with a long-duration main shock and shallow focus depth. The earthquake was in an area of the western Sichuan Basin, where the geologic environment is very fragile and the elevation gradient is very large. Consequently, there were numerous secondary geologic disasters, such as debris flows, flow-like landslides, and rock avalanches. Collapses and landslides were along the seismic zone. The numbers of fatalities, scale, and loss were unusual in world history (Table 7.1).

Experts and researchers from the Ministry of Land and Resources of the People's Republic of China conducted a detailed investigation of induced geo-disasters in Beichuan County, Shifang City, and Mianzhu City after the earthquake, finding some landslides and collapses. There were rolling stones throughout the hardest-hit areas, with average weight 20 t. Landslides and collapses caused road blockages and destroyed power infrastructure, communication stations, and other facilities. The landslides and rockfalls triggered by the earthquake blocked many rivers in Beichuan, Qingchuan, and Deyang in Sichuan Province, forming 34 dammed lakes. These included eight lakes with water volume more than 3 million m³, 11 with 1–3 million m³, and 15 with less than 1 million m³. Experts said that if these lakes were breached, serious flood damage would ensue and seriously threaten the downstream plain and power stations on the Minjiang River. Downstream areas would be completely submerged within 5 or 6 hrs of the breach, causing a flood disaster potentially as grave as that of the earthquake itself.

The large magnitude, long duration, and strong seismic response (1–2 g) with shallow focus depth caused the characteristics of the triggered geologic disasters to be very different from those with a common gravity environment. For example, the landslides induced by the Wenchuan earthquake frequently had high speeds and long runouts, thereby causing greater damage and casualties.

Case 1 Donghekou landslide in Qingchuan County

Among the flow-like landslides in Qingchuan County caused by the Wenchuan earthquake, the Donghekou landslide was extremely rapid with a long runout (Yin et al. 2009). This landslide was at the confluence of the Jinzhujiang River and smaller Hongshihe River (Fig. 7.1). Landslide material was mainly composed of

Table 7.1 Fatalities (≥ 30 people) from earthquake-triggered landslides (Yin et al. 2009)

No.	Name of the geo-disaster	Type of the geo-disaster	Location of the geo-disaster	Volume 10^4 m^3	Fatalities
1	Chengxi landslide	Slide	Wangjiayan, Old area of Beichuan County Town	480	1600
2	Yingtaogou landslide	Slide	Chayuanliang-cun Village, Chenjiaba Town, Beichuan County	188	906
3	Xinbei Middle School landslide	Slide	New area of New County Junior High School, Beichuan County	240	700
4	Jingjiashan rockfall	Rockfall	Main road of Southern Beichuan County Town	50	60
5	Hanjiashan landslide group	Slide	Team 1, Dujiaba-cun Village, Guixi Town, Beichuan County	30	50
6	Chenjiaba landslide	Slide	Chengjiaba Town, Beichuan County	1200	400
7	Hongyancun landslide	Slide	Hongyan-cun Village, Chenjiaba Town, Beichuan County	480	141
8	Taihongcun landslide	Slide	Taihong-cun Village, Chenjiaba Town, Beichuan County (landslide lake)	200	150
9	Donghekou landslide	Slide	Donghekou-cun Village, Hongguang Town, Qingchuan County	3000	780
10	Dayanke rockfall	Rockfall	Jianxin-cun Village, Quhe Town, Qingchuan County	70	41
11	Zhengjiashan landslide group	Slide	Xinping-cun Village, Nanba Town, Pingwu County	1250	60
12	Linjiaba landslide	Slide	Linjiaba Dam, Pingwu County	200	60
13	Maanshi landslide group	Slide	Maanshi-cun Village, Shuiguan Town, Pingwu County	400	34
14	Guantan landslide	Slide	Guantan-cun Village, Cuishui Town, An County	144	100
15	Yibadao-Xiaoguangjian landslide and rockfall	Slide Rockfall	Yanjiang Road, Mianyuan River, Mianzhu County	Area affected by numerous landslides and rockfalls	50

(continued)

Table 7.1 continued

No.	Name of the geo-disaster	Type of the geo-disaster	Location of the geo-disaster	Volume 10^4 m^3	Fatalities
16	Hongcun HPS landslide	Slide	Hongcun HPS, Shitingjiang River, Shifang County	100	150
17	Limingcun landslide	Slide	Liming-cun Village, Zipingpu Town, Dujiangyan City (National Road 213)	20	120
18	Xiaolongtan rockfall	Rockfall	Yinchanggou Resort, Pengzhou City	5.4	100
19	Dalongtan goukou rockfall	Rockfall	Yinchanggou Resort, Pengzhou City	10	100
20	Xiejiadianzi landslide	Slide	Team 7, Jiufeng-cun Village, Pengzhou City	400	100
21	Liangaiping landslide	Slide	Tuanshan-cun Village, Pengzhou City	40	30
22	Taian-9-team landslide group	Rockfall	Zhoujiaping, Qingchengshan Town, Dujiangyan City	3 landslides with total volume of 1.2 million m^3	62
23	Yingxiu-Wenchuan Road landslide and rockfall	Slide Rockfall	Along the Tourist Road from Dujiangyan to Jiuzhaigou Resort	Area affected by numerous landslides and rockfalls	1000
Total					6074

Fig. 7.1 Donghekou landslide in Qingchuan County (reprinted from Huang et al. (2012) with permission from Springer)

Fig. 7.2 Wangjiayan landslide in Beichuan County (reprinted from Huang et al. (2012) with permission from Springer)

carbonaceous slate, phyllite, and siliceous limestone (Wang et al. 2009). The landslide had a sliding distance of 2,400 m and volume of 10 million m^3 (Sun et al. 2009). At least 300 villagers were killed by the landslide, and two buses and a car were known to be buried by it.

Case 2 Wangjiayan landslide in Beichuan County

One of the most serious landslide disasters during the Wenchuan earthquake, the Wangjiayan in Beichuan County was also a typical flow-like landslide. It killed 1,600 people and destroyed hundreds of houses (Fig. 7.2). The landslide was only 300 m away from the rupture zone of a main central fault, and was composed of Cambrian sandstone, shale, and schist. The surface layer of the landslide was the accumulation of an ancient landslide. The landslide occurred on an anti-dip slope and had a volume of 4.8 million m^3. The height difference between the front and back edge was 350 m, with a sliding distance of 550 m (Yin et al. 2009).

Case 3 Tangjiashan landslide in Beichuan County

The Tangjiashan landslide occurred during the Wenchuan earthquake on the right bank of the Tongkou River, 6 km upstream from Beichuan County. This large landslide killed 84 people. It had a height difference of 650 m between the front

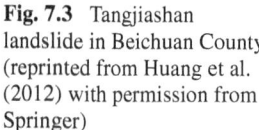

Fig. 7.3 Tangjiashan landslide in Beichuan County (reprinted from Huang et al. (2012) with permission from Springer)

and back edge (Fig. 7.3) and horizontal sliding distance of 900 m (Hu et al. 2009). The landslide formed an extremely large impounded lake of capacity 250 million m^3, and was composed of weathered silicates, sandstone, marlstone, and mudstone (Cui et al. 2009).

7.1.2 Current Research

Landslides and collapses are the main types of slope failure. Because of the action of a variety of internal and external geologic forces, the shapes, heights, incline angles, and stress states of slopes vary greatly. When the strength of slope cannot bear the stress distribution, deformation will occur and the slope may fail. Factors affecting slope stability fall into two categories, dominant and triggering. The dominant factors include the type and nature of geotechnical material, geologic structure of rock and soil, weathering, and groundwater activities. The triggering factors include precipitation, flowing water on the slope surface, earthquakes, and human activities (such as loading and artificial blasting). Among the factors, earthquakes are one of the most important triggers of landslides and numerous large-scale collapses. For example, in 25 August 1933 there was a large earthquake at the town of Diexi on the upper stream of the Minjiang River, which caused a large landslide and collapse that destroyed the town. The landslide and collapse blocked the river and formed a dammed lake called Diexi, with capacity 4–5 × 10^8 m^3. An earthquake measured at 8.5 Ms occurred in Chile in 1965, resulting in thousands of landslides and collapses. An investigation in southwestern Songpan and Pingwu counties by Chinese scholars revealed that secondary geo-disasters are common when the earthquake is Ms 7.0 and the slope incline exceeds 25°. The Wenchuan earthquake struck areas in the Longmenshan fault zone. The geologic structure there is active and the elevation difference is great, with a coverage of loose solid material. Strong vibration from the earthquake provoked instability of the mountain and a large number of collapses and landslides. This resulted in the accumulation of a large amount of loose solid material inside river channels, which could form debris flows under heavy rainfall.

When analyzing the problem of slope failure, investigators tend to focus on slope stability, failure development, and reinforcement methods. Most methods for analysis of slope failure are in the field of solid mechanics, such as the traditional limit equilibrium method, finite element method, and failure probability analysis (Duncan 1996). In addition, based on interdisciplinary research, some new methods have been applied to research on slope failure, such as fracture mechanics, dissipative structure theory, neural networks, artificial intelligence systems, stochastic simulation theory, large deformation finite element method, and static-dynamic analysis (Duncan 1996). These methods have produced achievements in the field of slope stability analysis and played a significant role in alleviating slope disasters. However, given the serious consequences of earthquake-induced slope failure, especially for the landslide problem, there should be studies beyond stability assessment by qualitative and quantitative methods. Dynamic behaviors should be investigated to obtain landslide destructive power, runout, and other parameters. These should be fully considered in seismic prevention design and reconstruction.

Since the 1960s, there have been landslides around the world with high sliding speeds and long runouts. For example, there have been sliding speeds in excess of 100 km/h and runouts to a few kilometers or even hundreds of kilometers (Devoli et al. 2008). These landslides behave somewhat like a liquid. This phenomenon multiplies the landslide flow velocity and distance, and expands the scope of the hazard to lives and property. Therefore, in recent years, research on the fluid character of flow-like landslides has emerged and made some progress. Kent (1966) proposed the theory of "trapped air induced fluid". Melosh (1979) introduced the acoustic field theory in fluid dynamics to interpret the failure of flow-like landslides. Hsu (1975) proposed the theory of non-cohesive granular flow. However, most research in this field has been limited to result analysis, theoretical assumptions, and inferences regarding the phenomenon. Numerical simulations of flow-like landslide motion and kinetic characteristic analyses have been rare.

To deal with this problem, a number of new and sophisticated numerical models have recently been developed, and there has been substantial modeling of the propagation stage. Discrete element methods, such as the distinct element method (DEM) and discontinuous deformation analysis (DDA), have been widely used to analyze the kinematic and runout behavior of discontinuous material such as blocky rock masses. For example, there have been dynamic DDA simulations of the Vaiont landslide, and the influence of geometric discontinuity on landslide kinematic behavior has been explored (Sitar et al. 2005). More recently, Utili and Crosta (2011) used a DEM model to simulate the evolution of natural cliffs subject to weathering. Li et al. (2012) presented a DEM model for simulating the movement of the Yangbaodi flow slide using a calibration-based approach. Additionally, computational fluid dynamics (CFD) has been used in analysis of slope failure caused by earthquakes (Sawada et al. 2004). In the field of CFD, SPH is a recently developed, mesh-free numerical method that originated as an astrophysics application (Lucy 1977; Liu and Liu 2003). The main advantage of SPH is that it bypasses the need for a numerical grid and avoids the severe mesh distortions caused by large deformation. It has been extended and applied to a vast range of situations. There have been a few preliminary applications of SPH to landslides, with some promising results. For example, SPH models combined with depth-integrated equations were developed and extended to simulate the propagation stage of flow-like landslides, debris flows, lahars, and avalanches (McDougall and Hungr 2004; Pastor et al. 2008, 2009; Haddad et al. 2010). These models have many unique features, such as an ability to account for nonhydrostatic and anisotropic internal stress states, material entrainment along the slide path, and rheology variation.

As described in previous chapters, our group established an SPH model based on earlier research work (Liu and Liu 2003; Moriguchi 2005; Nonoyama 2011) to simulate the post-failure motion of flow-like landslides caused by strong earthquakes (Huang et al. 2012). Software with a user-friendly interface has been developed (Huang et al. 2011). The model simulates landslide motion along a user-prescribed two-dimensional (2D) path. However, landslides travel across three-dimensional (3D) terrain. They may change direction, spread or contract, and split or join in response to local topography. For example, field investigations suggest that the dynamic behavior of the Donghekou landslide was controlled by geologic and tectonic conditions and local geomorphological aspects of the terrain (Sun et al. 2011). Therefore, 3D modeling is required to faithfully reproduce dynamic processes of flow-like landslides, from slide initiation to its cessation of motion. A large number of SPH particles should be used to simulate the large volume of failure material and complex 3D terrain. However, calculation efficiency is sharply reduced as the number of SPH particles increases. Therefore, parallel computing techniques should be incorporated into a 3D SPH model.

In this chapter, we describe a 3D SPH model based on our previous work that simulates earthquake-triggered flow-like landslide propagation across 3D terrain. For a Bingham fluid in 3D, the relationship between shear-strain rate and shear stress is

$$
\tau = \left(\eta + \frac{\tau_y}{\left(D_{\mathrm{II}} \right)^{1/2}} \right) D,
\tag{7.1}
$$

where τ is shear stress, η is the yield viscosity coefficient (characterizing the deformation resistance of the liquid), and D is the tensor of strain rates, which can be defined by

$$
D_{ij} = \frac{1}{2} \left(\frac{\partial u_i}{\partial x_j} + \frac{\partial u_j}{\partial x_i} \right).
\tag{7.2}
$$

D_{II} is the second invariant of the tensor of strain rates, and can be defined by

$$
D_{\mathrm{II}} = \frac{1}{2} D_{ij} D_{ij}.
\tag{7.3}
$$

The new numerical model was used to analyze propagation of the flow-like landslides triggered by the Wenchuan earthquake. The dynamic behavior of those landslides during propagation was determined, including the sliding path, maximum distance reached, flow velocities, and distribution and thickness of deposits. These factors are important in mapping hazardous areas and estimating hazard intensity, and for the identification and design of appropriate protective measures.

7.2 Donghekou Landslide

Qingchuan County is at the northern edge of the Sichuan Basin, at 104°36′–105°38′E and 32°12′–32°56′N. The county is at the junction of Sichuan, Gansu and Shanxi provinces, and is known as the "Golden Triangle". At the eastern part of the county are the Chaotian District, Shizhong District, Jiange County, and Shanxi Province. At the south are Jiangyou City, the west Pingwu County, and the north Gansu Province.

Qingchuan County is in the southwest part of the Qinling Mountains. Valleys in the county are usually "V" shaped. Strata in the county are Devonian and Silurian and they are distributed along the tectonic line. Rocks are mainly weathered phyllite, schist, sandstone, and limestone.

There are three regional active faults traversing Qingchuan County. The first is the Dujiangyan-Jiangyou fault through the southern part of the county. This fault had no signs of activity in the Wenchuan earthquake.

The second was the causative fault of the Wenchuan earthquake, called Yingxiu–Beichuan. This fault is divided into two branches in the county. The southern branch extends from Daduli to Xindianzi, Dalongchi, Liangshui, Chaba, Sanyuan, Guanyindian, Shengli, and Jindong. The northern branch is from Chaoyang (Hongmiaozi) to Guanyindian, Shiba, Guanzhuang, Maoba, and Dayuan. The branches merge at Guanyindian. In the Wenchuan earthquake, the two fractures showed signs of activity and ground ruptures, building damage, and geologic disasters occurred along them.

The third fault is called Pingwu–Qingchuan. It traverses the urban area of Qingchuan, Qingxi, Shazhou, Qingchuan County, Banqiao, Shangma, and Muyu (only 10–15 km from the Yingxiu–Beichuan fault). Although Pingwu–Qingchuan did not cause the Wenchuan earthquake, it was significantly affected by the Yingxiu-Beichuan fault and showed activity in the northern Qingchuan County. The main evidence was ground rupture, building damage, and geologic disaster along the fault in the northern part of the county. There were also several aftershocks along this fault.

In addition to the aforementioned faults, there were other faults in the county such as Jiujiaya, Qu River-Fangshi, Dashenhuo, Yaodu, and Tangjiahe.

The Wenchuan earthquake had a maximum intensity of 11° in the Chinese standard. According to statistics, 9,351 aftershocks were recorded after the earthquake, 193 of which were greater than Ms 4.0. The earthquake seriously affected

36 towns, with 4,695 people dead, 124 missing, and 15,390 injured. Seismic disasters impacted 31,833 households and 123,440 people. About 950,000 residential houses over 13,450,000 square meters collapsed, and 400,000 administrative buildings over 5 million square meters were destroyed. The most seriously affected areas were Magong, Shiba, Hongguang, Fangshi, Quhe, and Muyu. Casualties from the seismic disasters represented a large proportion of total deaths in the first five towns. For example, 62 of 88 people were killed by landslides and avalanches in the town of Quhe. In the town of Fangshi, these phenomena killed 22. More than 500 were killed by the seismic disasters in the towns of Hongguang and Magong.

Being representative of high-speed and long-runout landslides triggered by the Wenchuan earthquake, the Donghekou landslide in Qingchuan County was selected as a case study with site investigation, experimental research, and SPH simulation. The earthquake-induced geologic disasters in Qingchuan were severe, and it is necessary to study such disasters in this area to provide technical support for seismic design and reconstruction. However, the Donghekou landslide was an anti-dumping or horizontal structural slope. Under the action of a strong earthquake, the failure mechanism is unique, with loosening, fracture, or even disintegration over the entire slope along a certain slip surface.

The Donghekou landslide was at Donghekou village in Hongguang town. The village had 10 communities, 324 households, and 1,263 people. The landslide occurred at the confluence of the Qinzhujiang River and smaller Hongshihe River, and was mainly composed of carbonaceous slate, phyllite, and siliceous limestone. The landslide was surrounded by features that may have been caused by an air blast at the time of failure (Wang et al. 2009). The landslide had a sliding distance of 2,400 m, and a volume of 10 million m^3 (Sun et al. 2009). At least 300 villagers were killed by the landslide, and two buses and a car are known to have been buried. Two dammed lakes were formed by the landslide, Hongshihe and Donghekou (Figs. 7.4 and 7.5).

Fig. 7.4 Hongshihe dammed lake

Fig. 7.5 Donghekou dammed lake

7.2.1 Site Investigation

Site investigation of the Donghekou landslide was initially conducted to obtain landslide topography and support the SPH simulation.

A large number of landslides were induced by the Wenchuan earthquake, which made it difficult to complete geologic catalogs of affected areas. The traditional compilation technique requires a large number of on-site measurements. This requires substantial manpower and investment and threatens worker safety. Therefore, to improve the efficiency and accuracy of the investigation, advanced equipment and technology from the Chengdu University of Technology were used, which assured smooth progress of the project.

An advanced scanning system called the ILRIS-3D Intelligent Laser Ranging and Imaging System (Optech Inc., Ontario, Canada) was used for topography scanning of the Donghekou landslide. This 3D laser scanning system is the most advanced for acquiring 3D distance information of special multiple targets. The system can extend the point measurements of traditional systems to surface measurement. Thus, the scanning can deal with the scene in a complex environment and transfer a large number of 3D data directly to a computer. This enables reconstruction of a 3D model of the target, including point, line, surface, body, and other geometric data, which can be used for a variety of post-processing work. The scanning can be done in combination with a global positioning system (GPS) in field conditions, according to project needs. The result is position data of obvious locations and scene.

Table 7.2 lists the working parameters of the ILRIS-3D. There are some shortcomings of the system, such as limited laser power, scanning distance, and scope. Compared with traditional scanning technology, the ILRIS-3D can shorten the time of investigation and slope scanning.

3D data from on-site scanning after the earthquake were exported to a CAD system to construct a topographic map of the Donghekou landslide (Fig. 7.6).

Table 7.2 Working
parameters of ILRIS-3D

Model	ILRIS 3D
Manufacturer	Optech
Scanning distance/m	3 ~ 1000
Accuracy/mm·(100 m)$^{-1}$	±8
Scan trace	40° × 40°
Data sampling rate/(dot·s^{-1})	2000
Wave length of laser/nm	1500
Laser level	Class 1
Dimensions/cm^3	320 × 320 × 220
Machine mass/kg	13
Voltage/V	24
Software	Polyworks 8.0
RGB function	Yes

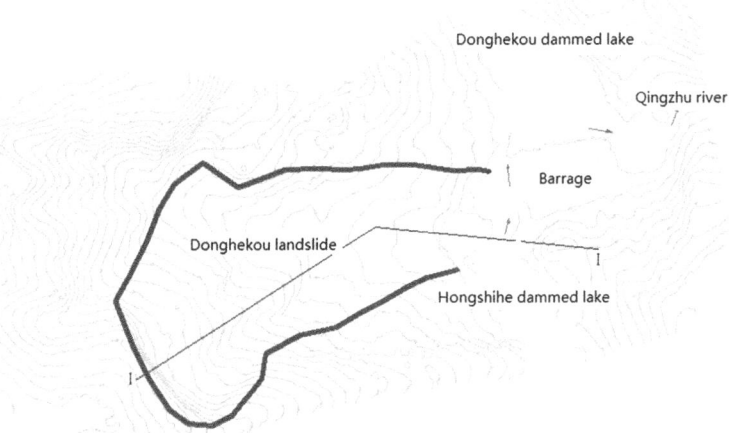

Fig. 7.6 Scan data of Donghekou landslide

In addition, the original slope and slip surface were obtained via integration of the 1:50000 topographic map of Qingchuan County prior to the earthquake (Fig. 7.7).

7.2.2 Experiments

① Experimental apparatus

We used a large-scale triaxial test system for coarse-grained soil (Fig. 7.8) at the State Key Laboratory of Geohazard Prevention and Geo-environment Protection (SKLGP), Chengdu University of Technology.

Fig. 7.7 Pre-earthquake topographic map of Qingchuan County

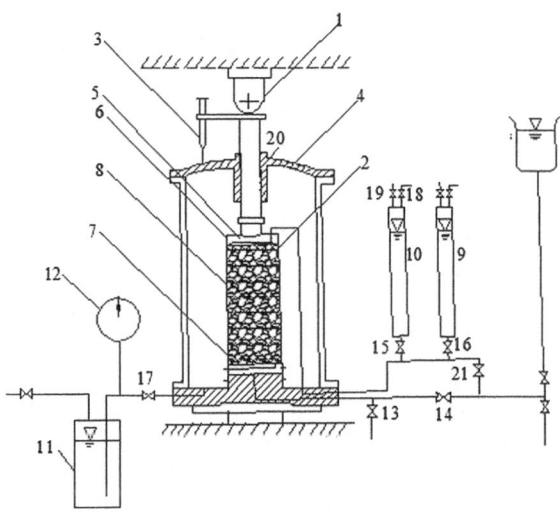

Fig. 7.8 Triaxial test system for coarse-grain soil. *1* dynamometer; *2* sample; *3* axial displacement meter; *4* pressure chamber cover; *5* top intake; *6* upper permeable plate; *7* lower permeable plate; *8* rubber membrane; *9* water measuring tube; *10* pressure reservoir; *11* pressure library; *12* pressure gauge; *13* pore pressure valve; *14* water inlet pipe valve; *15* drain valve; *16* water measuring valve; *17* confining pressure valve; *18* backpressure valve; *19* vent valve; *20* exhaust valve; *21* exhaust (water) valve

The experimental apparatus mainly consisted of the host, pumping system, axial hydraulic loading system, confining pressure system, counterpressure system, pore water pressure and volume change measuring system, and computer control system. The system is suitable for measuring shear strength and deformation of coarse-grain soil under the conditions that axial stress is not greater than 17 MPa (maximum axial load 1,200 kN) and confining pressure is not greater than 3.0 MPa.

② Experimental method

The undrained and unconsolidated (UU) shear test was used in the experiment. When the sample was saturated, the inlet and drain valves for water were closed. The confining pressure valve was opened to load confining pressure (a constant) on the soil sample. The value of the confining pressure should be based on actual loading of the landslide; it was set to the values 100, 200, 300, and 400 kPa.

Loading in the test was controlled by the strain rate, and loading speed was 1 mm/min. Given the limited test conditions, the confining pressure and counter pressure were applied by a manual control pane. Axial loading was recorded at every 0.5 mm of axial deformation. When the loading maximized, the test was continued until the axial strain exceeded 3–5 %. If there was no peak, the axial strain reached 15–20 %.

At the end of the test, axial pressure was removed before unloading the confining pressure. Then the exhaust vent and drain valve were opened for drainage and the pressure chamber cover was removed. After that, we wiped the remaining water around the sample, removed the rubber membrane, and cut the specimen for description. If necessary, water content and breakage of the soil sample were analyzed.

The triaxial test of coarse soil determines shear strength, and three or four samples under different confining pressure should be conducted (i.e., the minimum principal stress σ_3). The axial pressure is applied (i.e., principal stress difference $\sigma_1 - \sigma_3$) to destroy the soil sample, and shear-strength parameters cohesion (c) and angle of internal friction (φ) could be obtained according to Mohr–Coulomb theory.

③ Soil sampling and experimental study

Donghekou was a typical flow-like landslide with high speed and long runout. To study soil properties before failure, soil sampling was done on the western side of the landslide.

At first, on-site density measurements were made. To minimize error, there were three soil samples of different volumes for the measurement of mass, $20 \times 20 \times 20$ cm^3, $30 \times 30 \times 30$ cm^3, and $40 \times 40 \times 40$ cm^3. Sample mass can be divided by the volume to get the density (Fig. 7.9). The natural density of soil in the Donghekou landslide was 2,010 kg/m^3. The soil samples were transported to the SKLGP for moisture content and particle size testing.

Moisture content of soil material from the landslide was 12.70 %. The composition of particles in different size ranges is presented in Table 7.3 and Fig. 7.10.

④ Triaxial test and results

Four cases, with confining pressures 100, 200, 300, and 400 kPa, were considered in the triaxial test (Fig. 7.11). According to the Mohr–Coulomb strength theory and obtained stress state for the Donghekou landslide soil sample, c was 125 kPa and φ was 39° (Fig. 7.12).

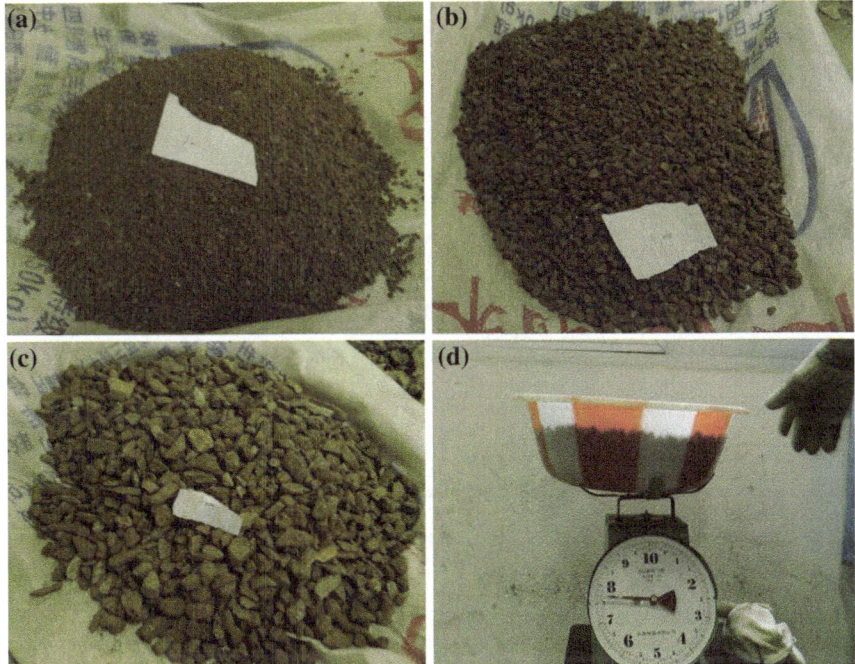

Fig. 7.9 Particle size test. **a** Φ=1mm, **b** Φ=5mm, **c** Φ=10mm, **d** weighting

Table 7.3 Particle composition of soil in Donghekou landslide

Particle size/mm	>60	60 ~ 40	40 ~ 20	20 ~ 10	10 ~ 5	5 ~ 2	2 ~ 1	<1
Mass/kg	8.81	2.06	10.72	13.25	13.06	16.54	4.73	16.77

Fig. 7.10 Pie chart of soil particle composition in Donghekou landslide

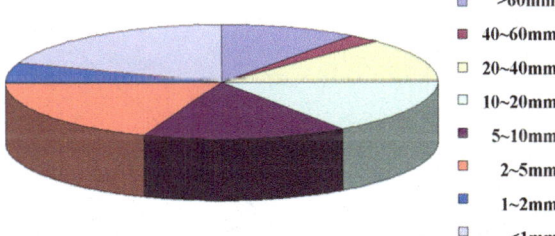

- >60mm
- 40~60mm
- 20~40mm
- 10~20mm
- 5~10mm
- 2~5mm
- 1~2mm
- <1mm

7.2.3 SPH Simulations

To reproduce the entire flow process of the Donghekou landslide, runout was analyzed with the SPH model. The landslide had several flow stages with long sliding distances. The runout analysis was for the main stage only. SPH

Fig. 7.11   Triaxial test. **a** fill, **b** compact, **c** test, **d** discharge.

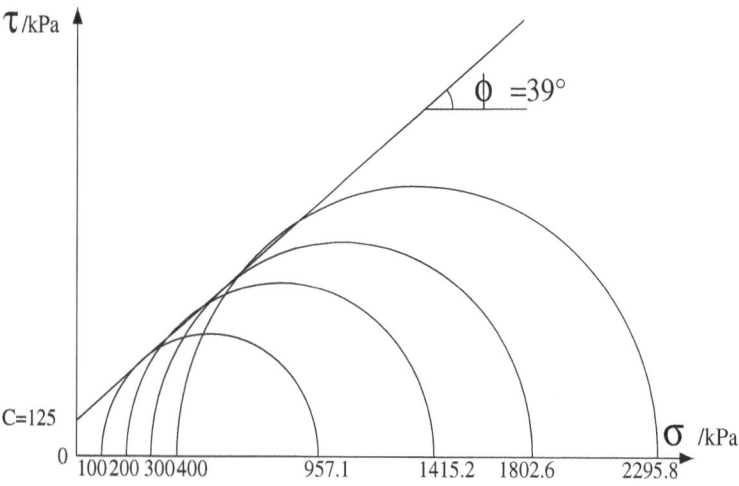

Fig. 7.12   Manual control panel for confining pressure

Table 7.4 Parameters in runout analysis of Donghekou landslide (reprinted from Dai and Huang (2014) with permission from Elsevier)

Density	ρ (kg/m^3)	2010
Equivalent viscosity coefficient	η (Pa·s)	2.0
Cohesion	c (kPa)	20.5
Angle of internal friction	ϕ (°)	39.0

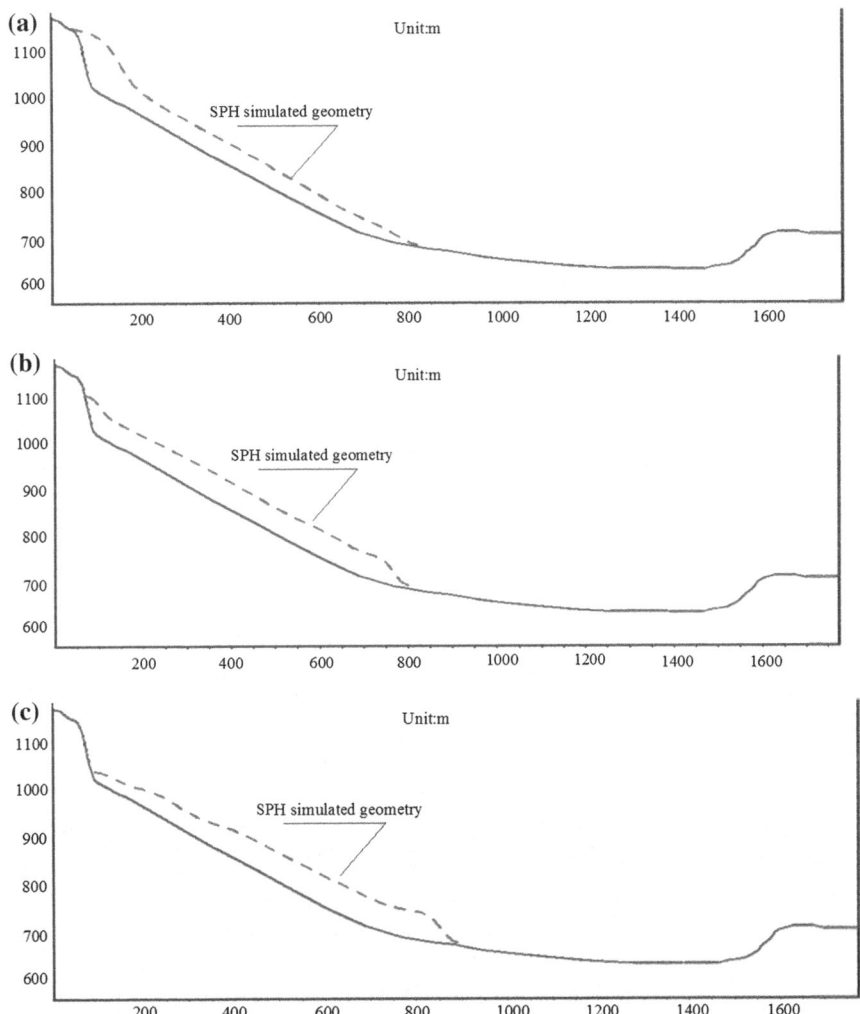

Fig. 7.13 Simulated runout process of Donghekou landslide (reprinted from Huang et al. (2012) with permission from Springer). **a** t = 0 s. **b** t = 10 s. **c** t = 20 s. **d** t = 40 s. **e** t = 60 s. **f** t = 80 s

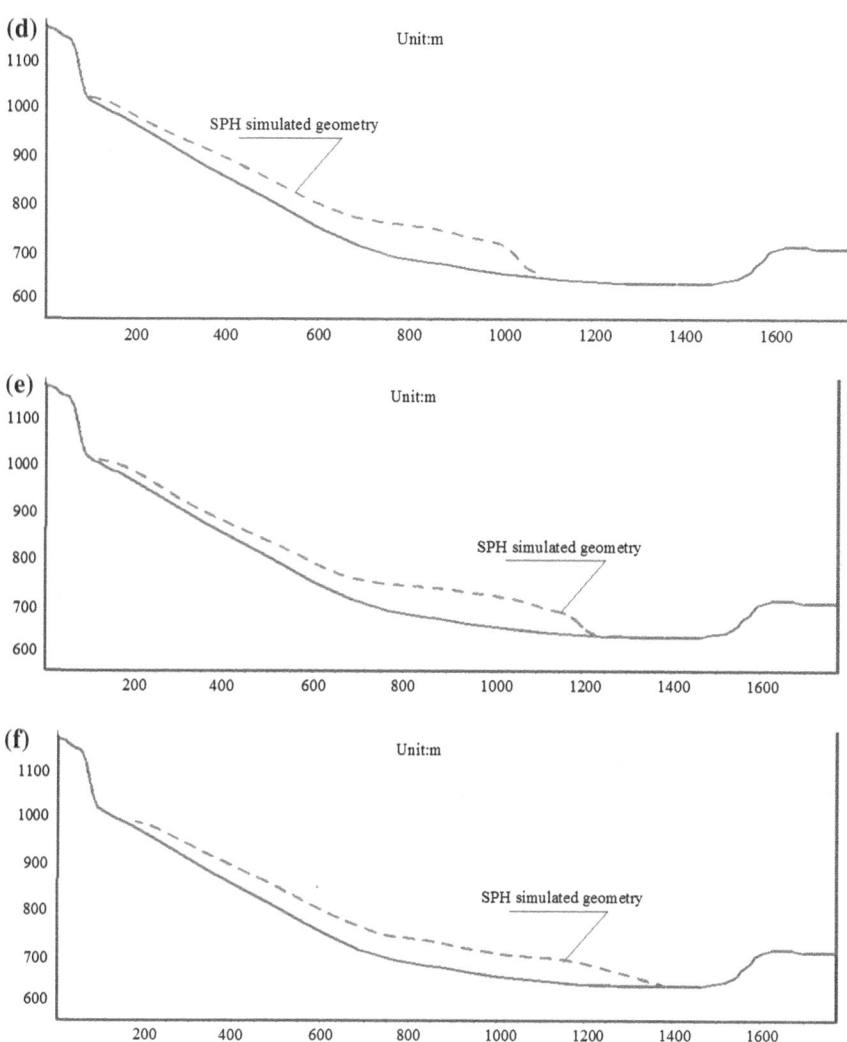

Fig. 7.13 continued

simulation parameters (Table 7.4) were derived from the site survey and tests. Figure 7.13 portrays the simulated runout process of the landslide.

This model simulates landslide motion along a user-prescribed 2D path. However, landslides travel across 3D terrain, and may change direction, spread or contract, and split or join in response to local topography. Field investigations suggested that the dynamic behavior of the Donghekou landslide was controlled by geologic and tectonic conditions and local geomorphological aspects of the terrain (Sun et al. 2011). Therefore, 3D modeling is required to faithfully reproduce the dynamic processes of flow-like landslides from slide initiation to its cessation of movement.

Using the 3D SPH-based numerical model, the propagation stage of the Donghekou landslide was predicted. Figure 7.14 shows positions of the avalanching mass with time. Time histories of the displacement and velocity (Figs. 7.15 and 7.16) indicate that the total sliding time of the landslide was about 100 s, and displacements of the landslide front and rear were 618 and 222 m, respectively.

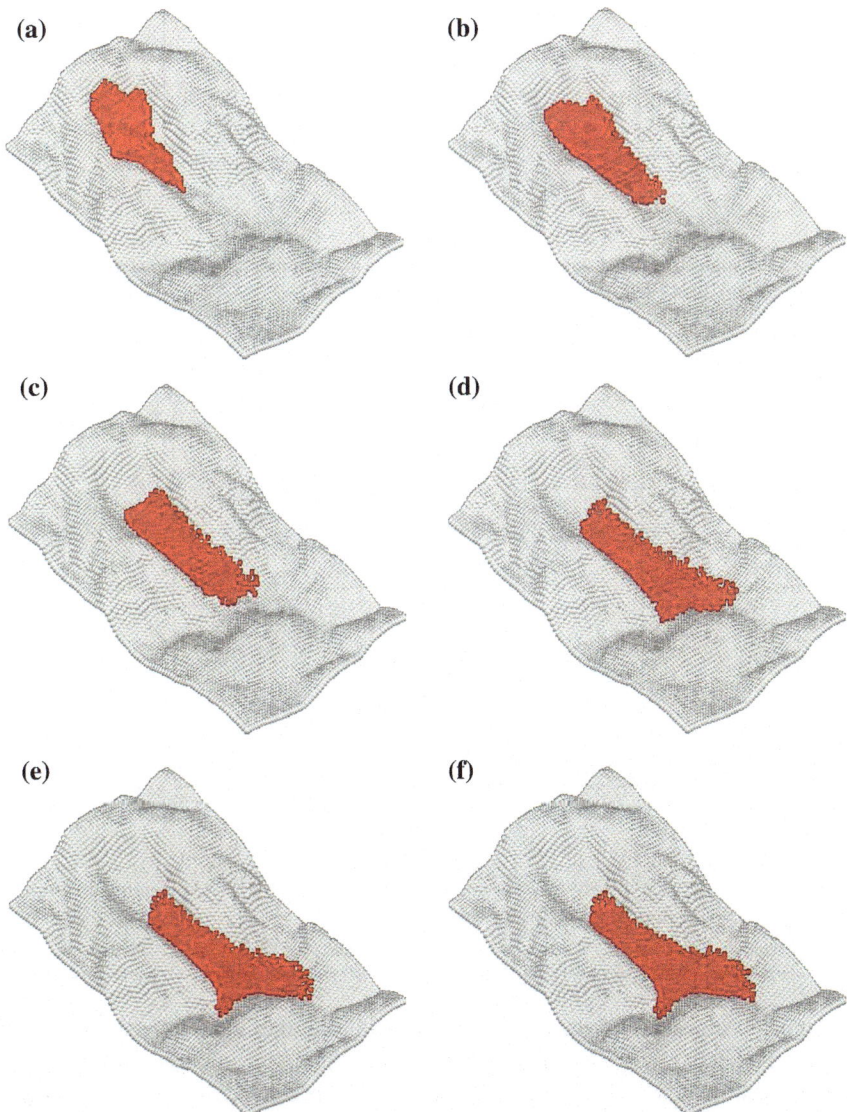

Fig. 7.14 Simulated propagation of Donghekou landslide (reprinted from Dai and Huang (2014) with permission from Elsevier). **a** t = 0 s. **b** t = 20 s. **c** t = 40 s. **d** t = 60 s. **e** t = 80 s. **f** t = 100 s

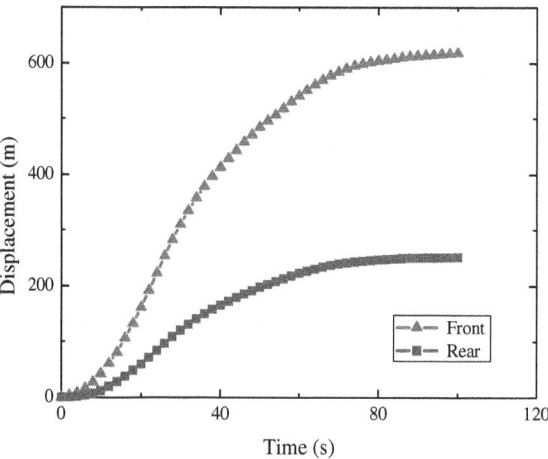

Fig. 7.15 Displacement time history for front and rear of Donghekou landslide (reprinted from Dai and Huang (2014) with permission from Elsevier)

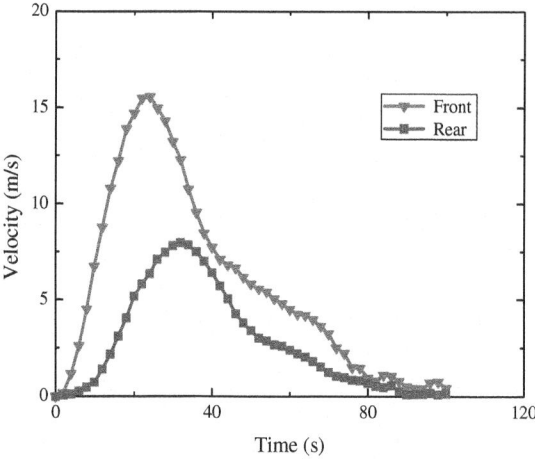

Fig. 7.16 Velocity time history for front and rear of Donghekou landslide (reprinted from Dai and Huang (2014) with permission from Elsevier)

Maximum velocities of the front and rear during the motion were 15.9 and 8.0 m/s, respectively, at 24 and 32 s after failure. Through comparison with the site survey data, there was a good agreement between predicted and observed shapes of the deposition zone (Fig. 7.17). From a cross section (Fig. 7.18) along line AB in Fig. 7.17, it is easy to see that final slide shapes simulated by both the 2D and 3D SPH models were very similar to surveyed landslide configurations.

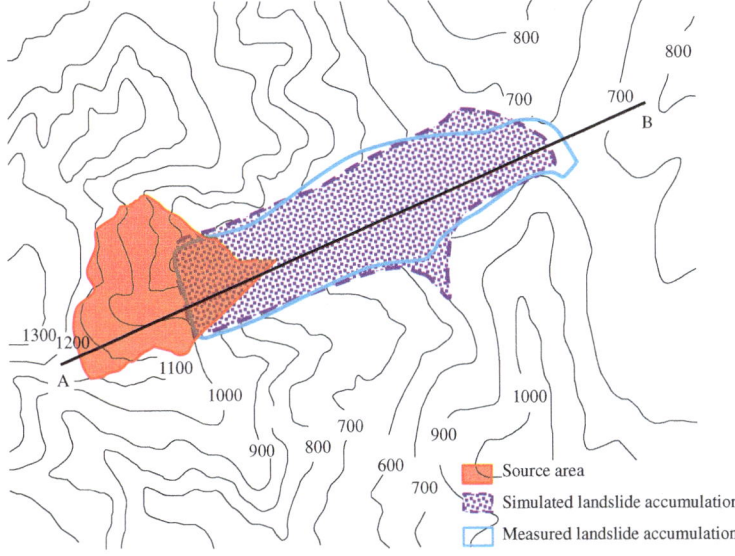

Fig. 7.17 Comparison of simulated and measured damage scope for deposition zone of Donghekou landslide (reprinted from Dai and Huang (2014) with permission from Elsevier)

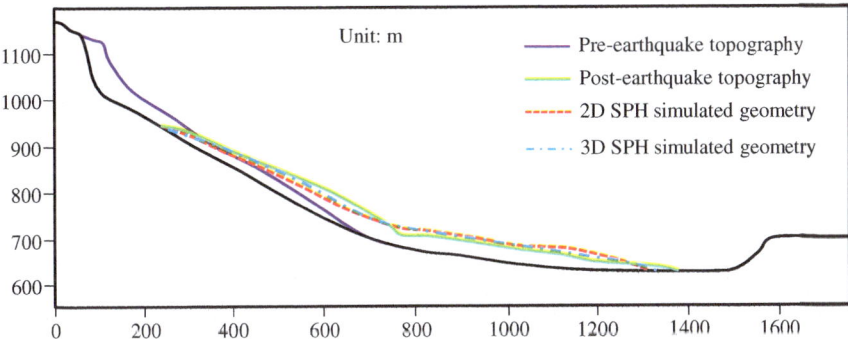

Fig. 7.18 Comparison of SPH simulation and survey data for Donghekou landslide, along line AB of Fig. 7.17 (reprinted from Dai and Huang (2014) with permission from Elsevier)

7.3 Tangjiashan Landslide

Tangjiashan Mountain is on the right bank of the Tongkou River, 6 km upstream from Beichuan County. During the Wenchuan earthquake, it failed and the resulting large landslide killed 84 people. The landslide was composed of weathered silicates, sandstone, marlstone, and mudstone (Cui et al. 2009). It had a height difference of 650 m between the front and back edge, and a horizontal sliding

Table 7.5 Parameters in runout analysis of Tangjiashan landslide (reprinted from Dai and Huang (2014) with permission from Elsevier)

Density	ρ (kg/m^3)	2000
Equivalent viscosity coefficient	η (Pa·s)	1.9
Cohesion	c (kPa)	30
Angle of internal friction	φ (°)	30.0

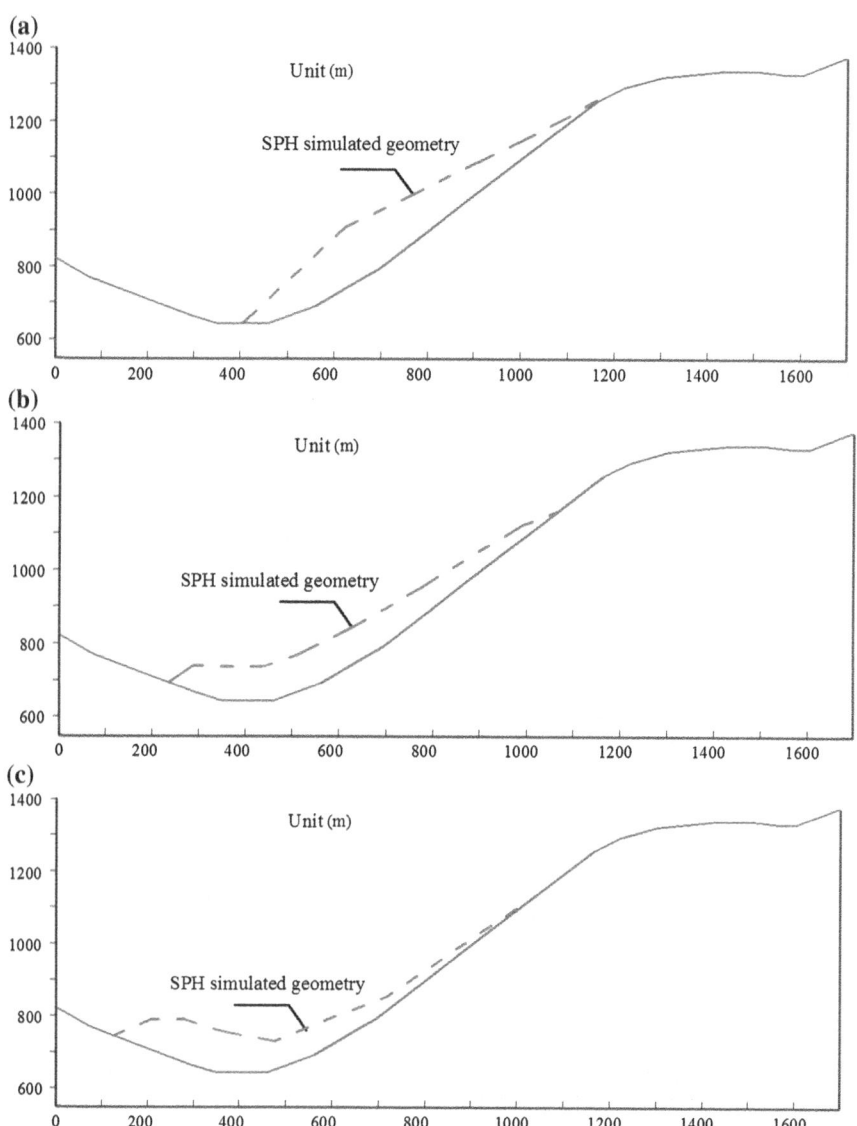

Fig. 7.19 Simulated runout process of Tangjiashan landslide (reprinted from Huang et al. (2012) with permission from Springer). **a** t = 0 s. **b** t = 6 s. **c** t = 12 s. **d** t = 18 s. **e** t = 24 s. **f** t = 30 s

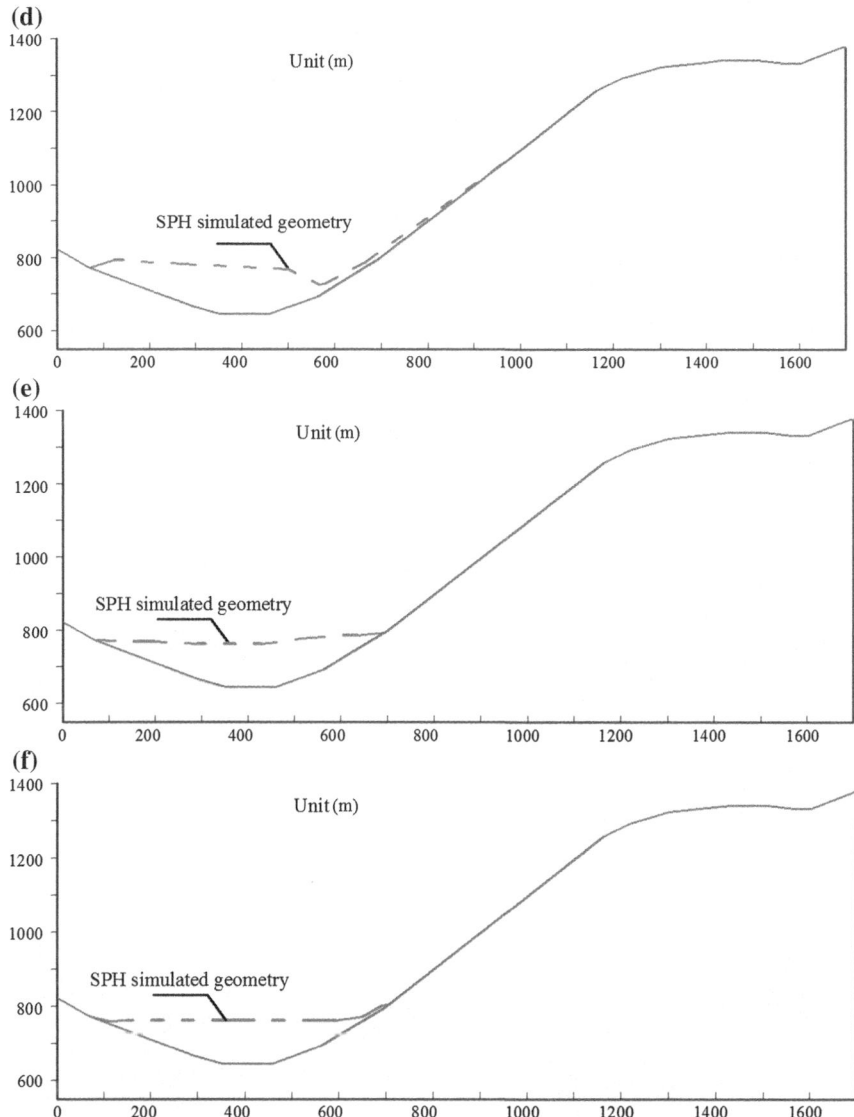

Fig. 7.19 continued

distance of 900 m (Hu et al. 2009). After failure, it formed an extremely large impounded lake with a capacity of 250 million m³.

We conducted 2D SPH simulations of the Tangjiashan landslide and studied the mechanisms that formed the impounded lake. Parameters in the runout analysis were based on Hu et al. (2009) (Table 7.5). Figure 7.19 presents the simulated runout process of the landslide and shows the evolution of the final slide shape.

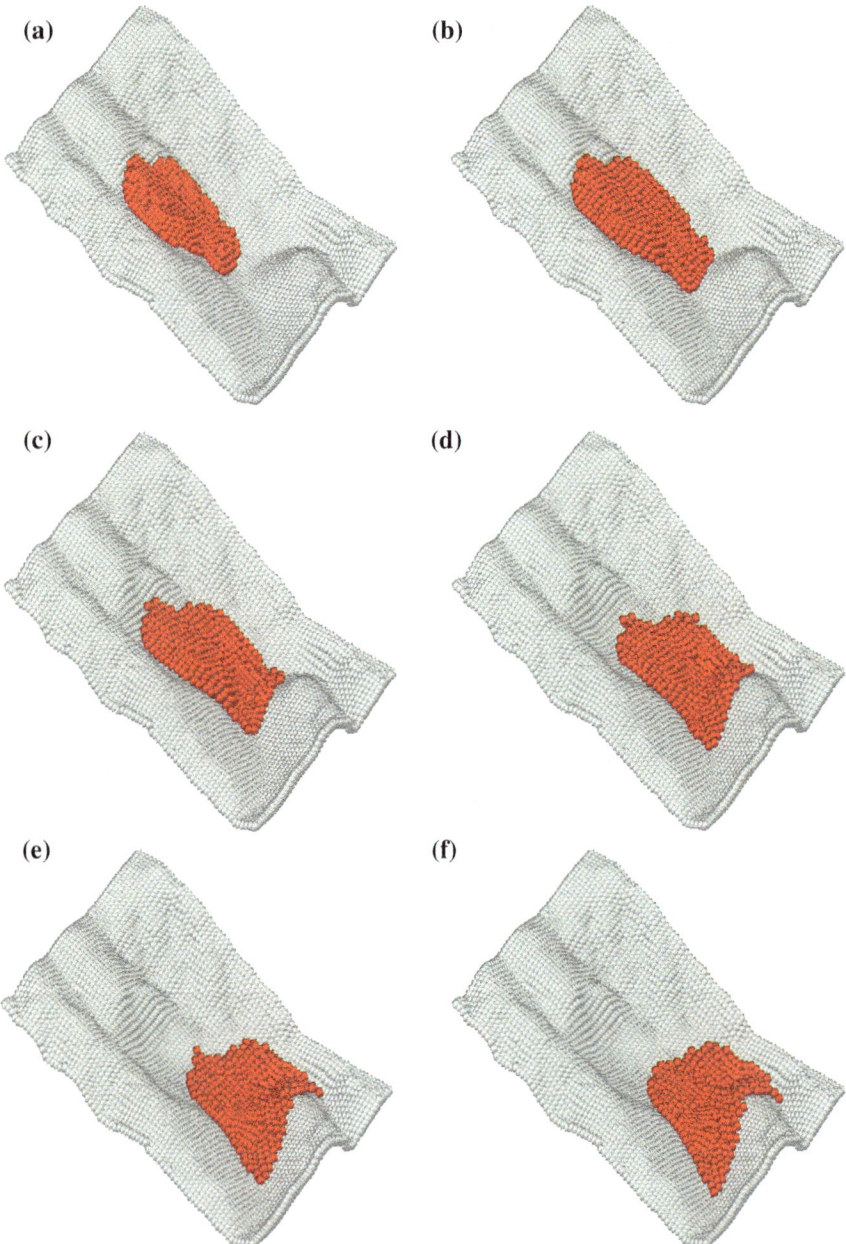

Fig. 7.20 Simulated propagation of Tangjiashan landslide (reprinted from Dai and Huang (2014) with permission from Elsevier). **a** t = 0 s. **b** t = 8 s. **c** t = 16 s. **d** t = 24 s. **e** t = 32 s. **f** t = 40 s

Besides the above runout analysis based on 2D SPH simulations, 3D numerical modeling of the Tangjiashan landslide propagation stage was performed using the 3D SPH model. In that model, soil material was discretized into a series of SPH particles of a certain diameter. The resulting model is shown in Fig. 7.20a. There were 10,747 particles in total, 2,569 for the sliding slope and 8,178 for the solid boundary, with particle spacing of 20 m. The number of particles versus depth varied with the depth of the sliding surface. To compare the simulated results with the 2D ones above, the parameters used for 3D simulation remained the same as in the 2D one. Figure 7.20 depicts the predicted propagation stage of the Tangjiashan landslide.

To show the dynamic characteristics of the soil material during sliding, displacement and velocity time histories of the landslide front and rear are presented in Figs. 7.21 and 7.22, respectively. At ~11 s after failure, the landslide

Fig. 7.21 Displacement time history for front and rear of Tangjiashan landslide (reprinted from Dai and Huang (2014) with permission from Elsevier)

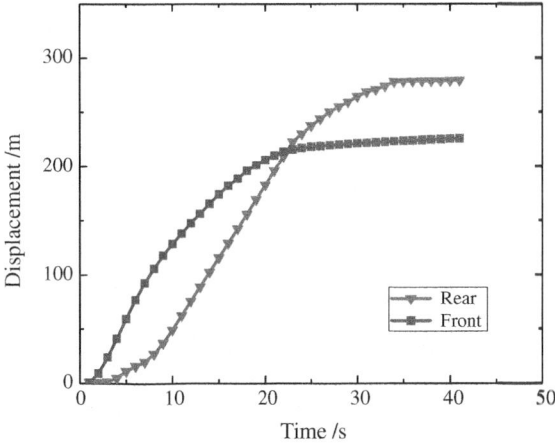

Fig. 7.22 Velocity time history for front and rear of Tangjiashan landslide (reprinted from Dai and Huang (2014) with permission from Elsevier)

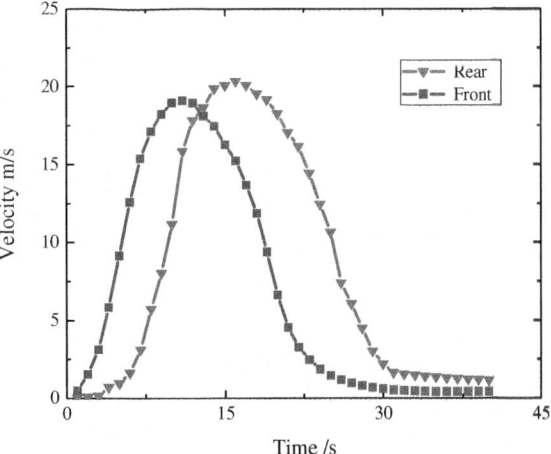

front reached a maximum velocity of 19.1 m/s; it then flowed across the Tongkou River, ascended the opposite slope and slowed. Afterward, the landslide changed its direction of motion and traveled along the river. After sliding for 30 s, the landslide gradually slowed to a stop and blocked the river. Total displacements of the front and rear were 225 and 280 m, respectively.

To demonstrate the accuracy of the SPH analysis for the Tangjiashan landslide, a comparison of the SPH-simulated and measured deposit zone is shown in Fig. 7.23. We see that the damage scope and accumulation extent were again modeled satisfactorily. A cross section of the landslide along line AB in the figure and comparison of the SPH-simulated geometry and surveyed landslide configuration are shown in Fig. 7.24. The green solid line is the pre-earthquake topography of the landslide. The purple solid line represents the post-earthquake topography, which was measured on-site by Hu et al. (2009). The yellow dot-dash line is the simulated geometry from the 3D SPH model developed here, and the red dashed line shows the 2D SPH simulation results. From the comparison, it is clear that the

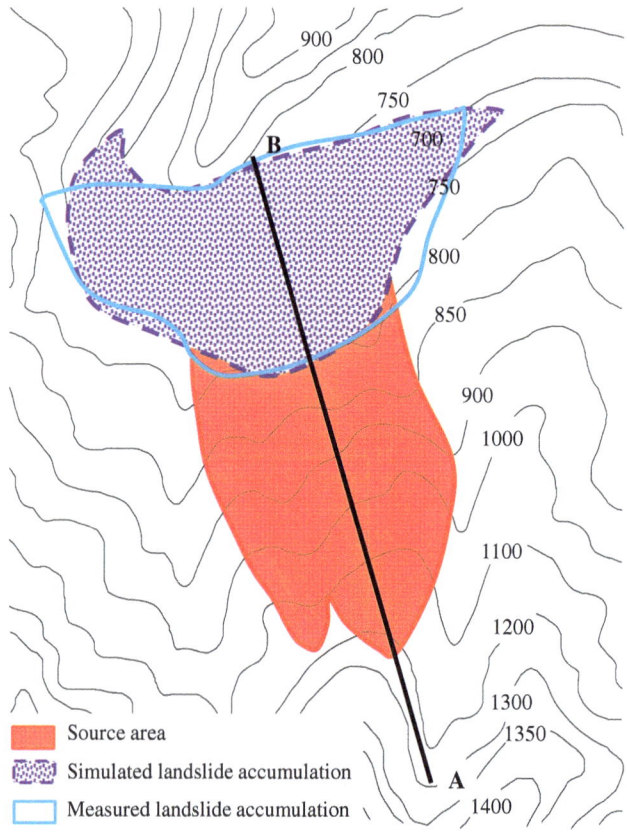

Fig. 7.23 Comparison of simulated and measured damage scope for deposition zone of Tangjiashan landslide (reprinted from Dai and Huang (2014) with permission from Elsevier)

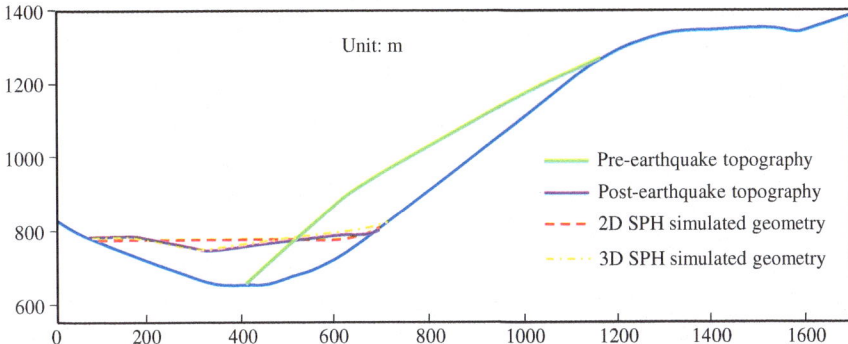

Fig. 7.24 Comparison of SPH simulation and survey data for Tangjiashan landslide, along line AB in Fig. 7.23 (reprinted from Dai and Huang (2014) with permission from Elsevier)

runout, slope coverage, and thickness from the 3D model are very similar to the post-earthquake topographic map. The 2D simulation showed slight deviation at the middle of the barrier dam. The main reason for this is that when the soil material reached the opposite slope, it traveled along the Tongkou River. The 2D model cannot simulate this phenomenon. In this model, all soil material was deposited in the riverbed, making it thicker than in the measured data.

7.4 Wangjiayan Landslide

The Wangjiayan landslide was in Beichuan County. It was a typical high-speed and long-runout landslide and was one of the most serious landslide disasters during the Wenchuan earthquake. It buried 1,600 people and destroyed hundreds of houses. The landslide was only 300 m from the rupture zone of the main central fault, and was composed of Cambrian sandstone, shale, and schist. The surface layer of the landslide was an accumulation of an ancient landslide. On an anti-dip slope, the landslide had a volume of 4.8 million m³. The height difference between the front and back edge was 350 m, with a sliding distance of 550 m (Yin et al. 2009).

Simulation parameters are shown in Table 7.6, with values derived from local engineering experience. Figure 7.25 depicts the simulated runout process of the landslide.

Table 7.6 Parameters in runout analysis of Wangjiayan landslide (reprinted from Dai and Huang (2014) with permission from Elsevier)

Density	ρ (kg/m³)	2000
Equivalent viscosity coefficient	η (Pa·s)	1.9
Cohesion	c (kPa)	30
Angle of internal friction	φ (°)	30.0

We conducted a 3D SPH simulation of Wangjiayan landslide propagation. Parameters were the same as those in the 2D simulation. Figure 7.26 presents the predicted propagation stage of the landslide, which slid rapidly down the slope during the earthquake and then spread out in all directions at the foot of the

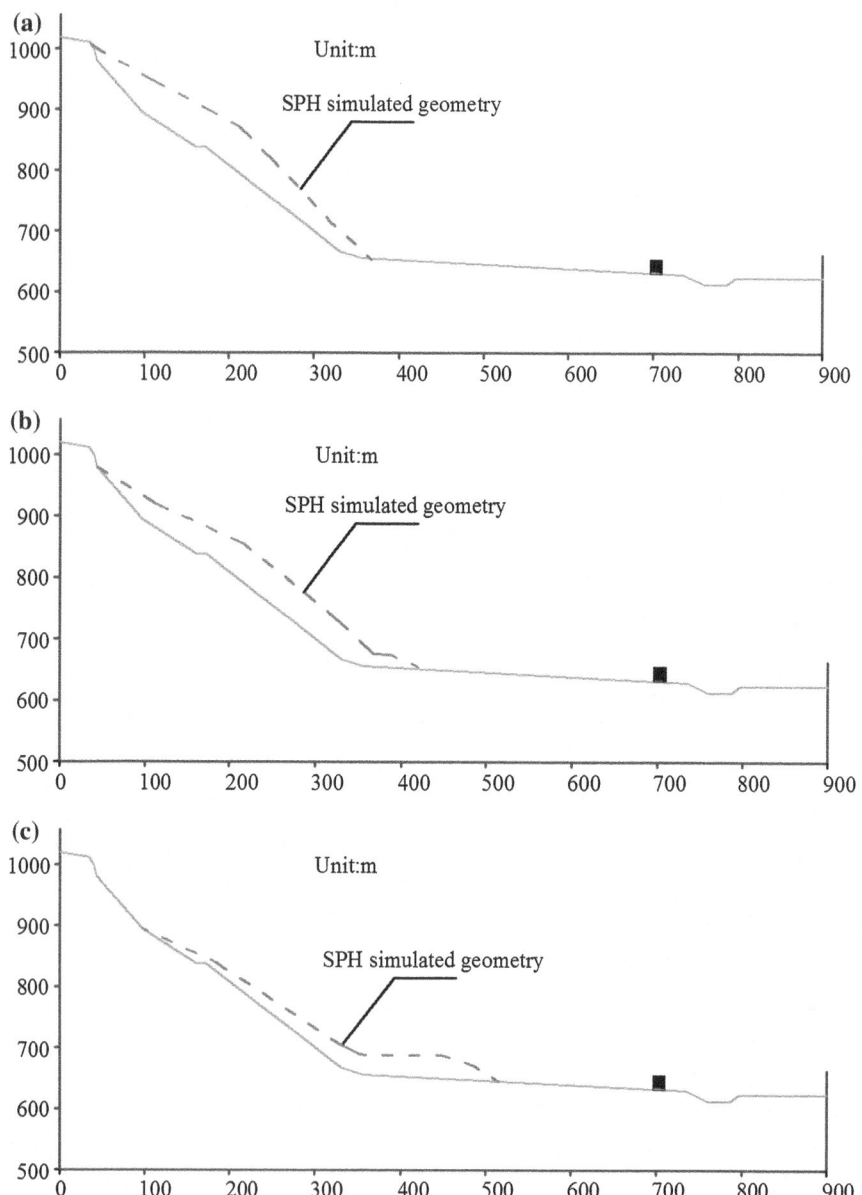

Fig. 7.25 Simulated runout process of Wangjiayan landslide (reprinted from Huang et al. (2012) with permission from Springer). **a** t = 0 s. **b** t = 6 s. **c** t = 12 s. **d** t = 18 s. **e** t = 24 s. **f** t = 30 s

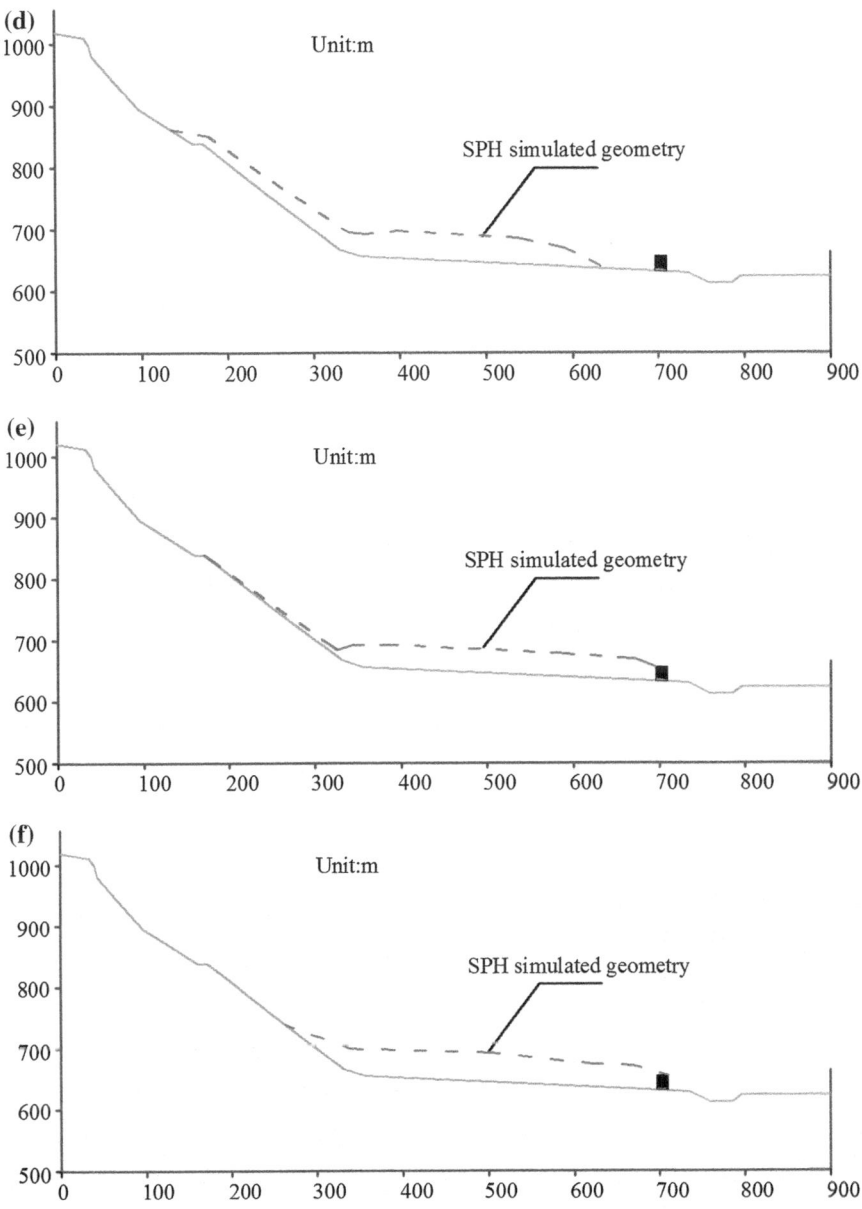

Fig. 7.25 continued

mountain. Time histories of displacement and velocity (Figs. 7.27 and 7.28) indicate that the entire landslide duration was around 30 s. Maximum velocities of the landslide front and rear were 23.9 and 16.0 m/s, and final displacements were 296 and 191 m, respectively. The shape of the simulated deposition zone matches the

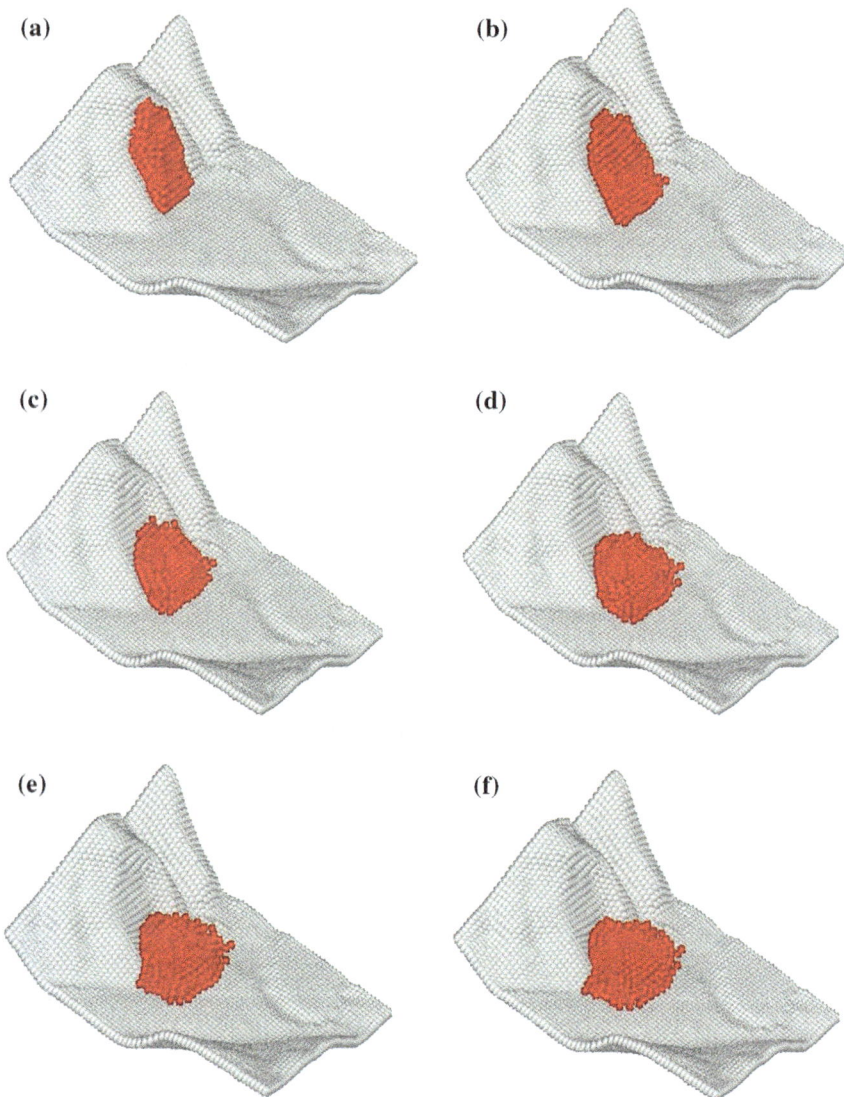

Fig. 7.26 Simulated propagation of Wangjiayan landslide (reprinted from Dai and Huang (2014) with permission from Elsevier). **a** t = 0 s. **b** t = 8 s. **c** t = 16 s. **d** t = 24 s. **e** t = 32 s. **f** t = 40 s

observed one well (Fig. 7.29). A cross section of the landslide is along line AOB in Fig. 7.29, for comparison of the SPH-simulated geometry with the measured landslide configuration (Fig. 7.30). The final simulated slide shape fits the measured post-earthquake topography well. The runout simulated by the 2D model is slightly larger, since soil material can only flow along a prescribed 2D path without spreading.

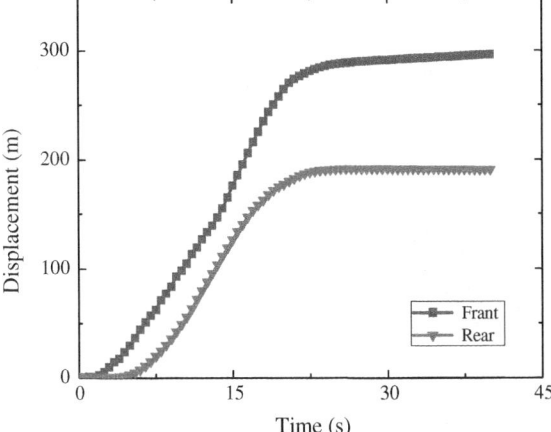

Fig. 7.27 Displacement time history for front and rear of Wangjiayan landslide (reprinted from Dai and Huang (2014) with permission from Elsevier)

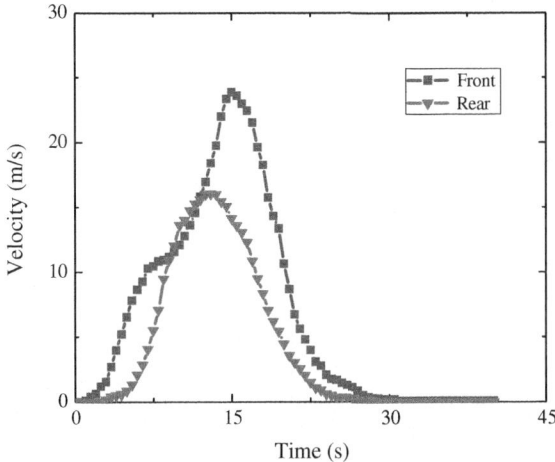

Fig. 7.28 Velocity time history for front and rear of Wangjiayan landslide (reprinted from Dai and Huang (2014) with permission from Elsevier)

From the simulated results shown above, it can reasonably be concluded that the 3D SPH model captures entire dynamic processes from slide initiation to its cessation of motion. The stage of impact and redirection caused by the complex 3D terrain can be determined, which was not so for the previous 2D model. Simulated shapes of deposition zones can be used in combination with sliding paths to map hazardous areas, which will be important in engineering risk analyses and management.

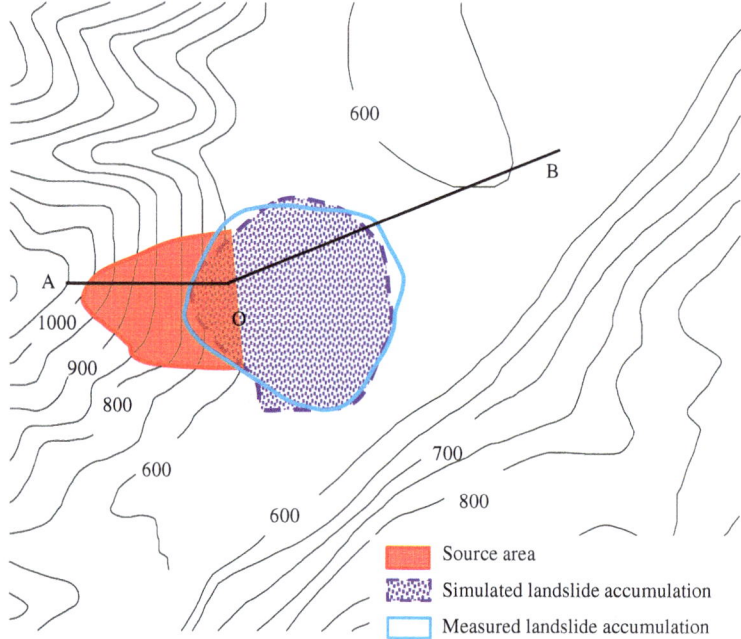

Fig. 7.29 Comparison of simulated and measured damage scope for the deposition zone of the Wangjiayan landslide (reprinted from Dai and Huang (2014) with permission from Elsevier)

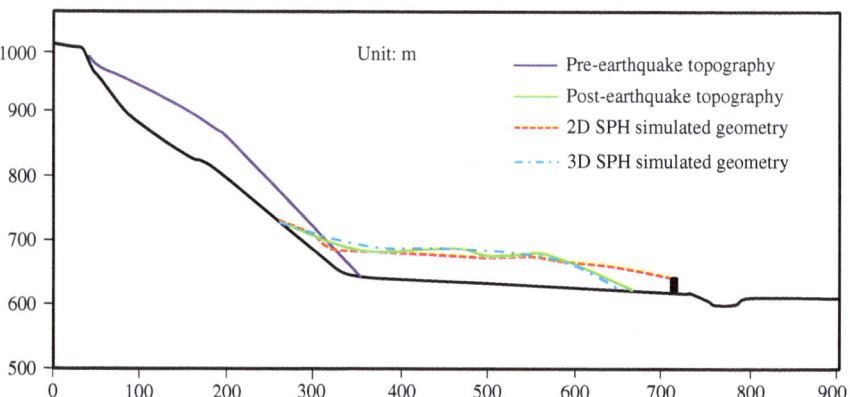

Fig. 7.30 Comparison of SPH simulation and survey data for the Wangjiayan landslide, along line AOB in Fig. 7.29 (reprinted from Dai and Huang (2014) with permission from Elsevier)

7.5 Summary

The Ms 8.0 Wenchuan earthquake of 2008 triggered a large number of flow-like landslides and other types of geologic hazards, which caused great damage and numerous casualties. During reconstruction after the earthquake, more attention should be given to these landslides, to identify flow mechanisms and investigate specific characteristics that could improve hazard assessment and facilitate reconstruction. The runout analyses and propagation prediction herein were targeted at the flow-like landslide induced by the Wenchuan earthquake. The results can contribute to post-earthquake reconstruction.

For the Donghekou landslide in Qingchuan County, through on-site investigation and related experiments, shear-strength parameters of soil samples were obtained. These furnish data support to reveal the flow mechanism of flow-like landslides.

Runout analyses of the Tangjiashan, Wangjiayan, and Donghekou landslides, which were typical flow-like landslides triggered by the Wenchuan earthquake, were conducted as applications of the SPH technique to actual flow-like landslides. From comparisons of SPH results and surveyed configurations of the three landslides, SPH-simulated geometries were very similar to the surveyed configurations. This indicates that the simulations could accurately reproduce entire flow processes associated with typical earthquake-induced, flow-like landslides, and provide an effective tool for investigating landslide flow mechanisms. Essential landslide characterization parameters, including runout and velocity, and other fundamental dynamic behaviors mentioned above, can be derived from SPH simulation.

As shown by the verification and validation above, the SPH numerical model can theoretically reproduce the entire propagation stage of geomaterials. In addition to sliding velocity, slope configuration, and deposition zone shape, other fundamental dynamic behaviors can be determined from SPH analysis. These include the impact force in numerical simulation of flow-like landslides. Based on these dynamic behaviors, hazard assessments can be put into practice and help decision makers improve risk management and disaster prevention. Practically, however, it was nearly impossible during the Wenchuan earthquake to record specific real-time impact forces and velocities of landslides in mountainous areas, and no relevant field data were available. Except for slope configurations that can be surveyed post-earthquake, there is still a lack of co-seismic response data on entire propagation stages of the landslides. Therefore, our simulations of flow-like landslides were mainly based on comparisons of pre- and post-earthquake terrain data (runout, slope configuration, and shapes of deposition zones) from a few actual landslides.

The SPH simulations described in this chapter remain inadequate, and further calibration and validation are required for the entire flow processes of the studied landslides. However, improved simulation of landslide materials is in progress. Moreover, the authors are convinced that the present simulation results can still support geologic hazard assessments. They can also help select suitable locations for post-disaster reconstruction and avoid the devastating impacts of landslides on various engineering structures.

References

Cui, P., Zhu, Y. Y., Han, Y. S., Chen, X. Q., & Zhuang, J. Q. (2009). The 12 May Wenchuan earthquake-induced landslide lakes: Distribution and preliminary risk evaluation. *Landslides, 6*(3), 209–223.

Dai, Z. L., & Huang, Y. (2014). A three-dimensional model for flow slides in municipal solid waste landfills: the sequel of an original model. *Engineering Geology.* doi:10.1016/j.enggeo.2014.03.018.

Devoli, G., Blasio, F. V., Elverhøi, A., & Høeg, K. (2008). Statistical analysis of landslide events in Central America and their run-out distance. *Geotechnical and Geological Engineering, 27*(1), 23–42.

Duncan, J. M. (1996). State of the art: Limit equilibrium and finite-element analysis of slopes. *Journal of Geotechnical Engineering, ASCE, 122*(7), 577–596.

Haddad, B., Pastor, M., Palacios, D., & Munoz-Salinas, E. (2010). A SPH depth integrated model for Popocatepetl 2001 lahar (Mexico): Sensitivity analysis and runout simulation. *Engineering Geology, 114*(3–4), 312–329.

Hsu, K. J. (1975). Catastrophic debris streams (Sturzstroms) generated by rockfalls. *Geological Society of America Bulletin, 8,* 225–256.

Hu, X. W., Huang, R. Q., Shi, Y. B., Lv, X. P., Zhu, H. Y., & Wang, X. R. (2009). Analysis of blocking river mechanism of Tangjiashan landslide and dam-breaking mode of its barrier dam. *Chinese Journal of Rock Mechanics and Engineering, 28*(1), 181–189. (in Chinese).

Huang, R. Q., & Li, W. L. (2009). Analysis of the geo-disasters triggered by the 12 May 2008 Wenchuan Earthquake, China. *Bulletin of Engineering Geology and the Environment, 68*(3), 363–371.

Huang, Y., Dai, Z. L., Zhang, W. J., & Chen, Z. Y. (2011). Visual simulation of landslide fluidized movement based on smoothed particle hydrodynamics. *Natural Hazards, 59*(3), 1225–1238.

Huang, Y., Zhang, W. J., Xu, Q., Xie, P., & Hao, L. (2012). Run-out analysis of flow-like landslides triggered by the Ms 8.0 2008 Wenchuan earthquake using smoothed particle hydrodynamics. *Landslides, 9*(2), 275–283.

Kent, P. E. (1966). The transport mechanism in catastrophic rock falls. *Journal of Geology, 74,* 79–83.

Li, W. C., Li, H. J., Dai, F. C., & Lee, L. M. (2012). Discrete element modeling of a rainfall-induced flowslide. *Engineering Geology, 149,* 22–34.

Liu, G. R., & Liu, M. B. (2003). *Smoothed particle hydrodynamics: A mesh-free particle method.* Singapore: World Scientific Press.

Lucy, L. B. (1977). A numerical approach to the testing of the fission hypothesis. *Astronomical Journal, 82*(12), 1013–1024.

McDougall, S., & Hungr, O. (2004). A model for the analysis of rapid landslide motion across three-dimensional terrain. *Canadian Geotechnical Journal, 41*(6), 1084–1097.

Melosh, H. J. (1979). Acoustic fluidization: A new geologic process. *Journal of Geophysics, 84,* 7513–7520.

Moriguchi, S. (2005). CIP-based numerical analysis for large deformation of geomaterials. *Ph.D. Dissertation*, Gifu University, Japan.

Nonoyama, H. (2011). Numerical application of SPH Method for deformation, failure and flow problems of geomaterials. PhD thesis, Gifu University, Japan.

Pastor, M., Haddad, B., Sorbino, G., Cuomo, S., & Drempetic, V. (2008). A depth-integrated, coupled SPH model for flow-like landslides and related phenomena. *International Journal for Numerical and Analytical Methods in Geomechanics, 33*(2), 143–172.

Pastor, M., Herreros, I., Merodo, J. A. F., Mira, P., Haddad, B., Quecedo, M., et al. (2009). Modelling of fast catastrophic landslides and impulse waves induced by them in fjords, lakes and reservoirs. *Engineering Geology, 109*(1–2), 124–134.

Sawada, K., Moriguchi, S., Yashima, A., Zhang, F., & Uzuoka, R. (2004). Large deformation analysis in geomechanics using CIP method. *JSME International Journal Series B: Fluids and Thermal Engineering, 47*(4), 735–743.

Sitar, N., MacLaughlin, M. M., & Doolin, D. M. (2005). Influence of kinematics on landslide mobility and failure mode. *Journal of Geotechnical and Geoenvironmental Engineering, 131*(6), 716–728.

Sun, P., Zhang, Y. S., & Shi, J. S. (2009). Analysis on the dynamical process of Donghekou rapid long runout landslide-debris flow triggered by great Wenchuan earthquake. In *Proceedings of the International Symposium and the 7th Asian Regional Conference of IAEG* (pp. 902–909), September 9–11, Chengdu, China.

Sun, P., Zhang, Y. S., Shi, J. S., & Chen, L. W. (2011). Analysis on the dynamical process of Donghekou rockslide-debris flow triggered by 5.12 Wenchuan earthquake. *Journal of Mountain Science, 8*(2), 140–148.

Utili, S., & Crosta, G. B. (2011). Modelling the evolution of natural cliffs subject to weathering: 2. Discrete element approach. *Journal of Geophysical Research, 116*, F01017.

Wang, F. W., Cheng, Q. G., Highland, L., Miyajima, M., Wang, H. B., & Yan, C. G. (2009). Preliminary investigation of some large landslides triggered by the 2008 Wenchuan earthquake, Sichuan Province, China. *Landslide, 6*(1), 47–54.

Yin, Y. P., Wang, F. W., & Sun, P. (2009). Landslide hazards triggered by the 2008 Wenchuan earthquake, Sichuan, China. *Landslide, 6*(2), 139–152.

Printed by Printforce, the Netherlands